THE
DELIVERANCE

THE DELIVERANCE

Paul Zecos

Copyright © 2015 by Paul Zecos

Second Edition

Contents

1. The Correct Education

Reading, Math and writing are the means of education that schools must teach but what is the purpose of education?

The questions that a correct education must answer are: What is your purpose in life? What is right? How to think rightly? How to rightly and/or correctly survive, thrive, help others and be Joyful in Peace?

To survive, thrive, help others and fully enjoy living within the current democratic-capitalistic systems one needs to **at least** understand correctly and coherently the foundational forces of **physics, chemistry, biology, psychology, business, economics and politics;** and one must understand **the causation and correlations between them.**

The foundational forces (factors,) causation and correlations of all these academic disciplines are explained simply in the Education Exhibits below.
Don't you have to use every day some basic understanding of how each of these fields of knowledge work?
Does anyone deny that education in each and all of these academic disciplines is necessary and critically important for survival, prosperity and well- being in this day and age?

Is there any academic discipline which shows or even looks to find the causation, correlations and connections among all the academic disciplines? No.

So, if those in academia, who have disagreements and opposing views within each of these disciplines— as shown by their further specializations— don't teach the causation and correlations among the subjects they teach and they are not even looking systematically for those causation and/or correlations, so that the totality of what they teach is disconnected, discombobulated, self-contradicted, confusing, incoherent and doesn't make sense, why do you call them teachers and Professors?

How can one expect children to not lose their integrity, "wholeness," oneness while they are taught so much that is self-contradicted and disintegrated?

Given that the current education that our children get at school and College is disconnected, discombobulated, self-contradicted, confusing, incoherent, and doesn't make sense shouldn't the education be at least by interesting and fun methods rather than by enforcement through bossing children and by multi-hour babel of required homework? Then, testing and ranking them like dogs and making those who don't work hard to complete useless babel feel dumb, is a brilliant way for producing a dumb future for them!

There was a wonderful educator on TV who discussed alternatives on education that are not based on a teacher teaching and the kids listening and doing homework (even though this should not be excluded either) but on kids choosing what they want to learn about and having an interactive environment with older and younger kids and a teacher in which they can search, have dialogues, experiment and learn about it.

There should certainly be a separate academic discipline (or as part of philosophy) for high schools and Colleges that discusses the foundational forces(factors,) within physics, chemistry, biology, psychology, business, economics and politics and the causation and correlations between them.

The following Unified Theory integrates all these Academic disciplines. It proves that there is an opposition— actually there are two oppositions at the same time— one within and one between every academic discipline, everything and everyone and it is founded on the physics of Nature.

Those two oppositions are correctly and accurately described as a Cross.

If you believe me, please skip the particular Exhibits or the content that may seem too technical at this time for your interest.

EDUCATION

EXHIBITS

1. Thermodynamics

We will start with thermodynamics because that is what I was studying as part of Chemical Engineering when the cops started chasing me on the streets of Athens, because we asked for free speech.
By running while in hot pursuit by a bunch of crazy cops I had to use and experience all the thermodynamics that I knew to escape and as a result thermodynamics are strongly etched in my brain.

Everything may transform from one state of matter to the other with increasing temperatures, from solid to liquid to gas to plasma.
We describe these 4 fundamental states of all matter on a 4-axis Cross with each axis representing a different state of matter.

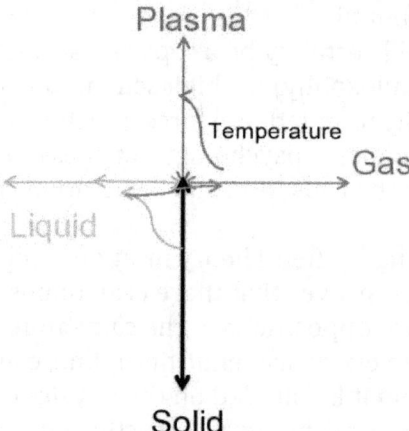

The four states of matter are very different from each other in terms of how they occupy space and how they move in space over time.
-Solids, because they don't change their shape, can move only in one direction at a time.
-Liquids can move in one (up-down) and/or in two dimensions at a time as moving surfaces and in multiple directions at the same time.
-Gases move in all 3 space dimensions at a time, shrinking in volume and expanding in all 3 space dimensions as temperatures change and occupying whatever volume is available.
-Plasma moves in 1, 2 or 3 space dimensions at a time: the rays of light from the plasma travel in 1 dimension at a time, the plasma moves as a surface in 2 Dimensions; and the heat from it spreads in all 3 space Dimensions synchronously; (i.e. at the same time.)

The direction of motion of solids is deterministic, meaning that it can be predicted how they will move, exactly. That is why we can predict at exactly what time (at a particular location) the Sun will rise in a week, month or a year.
In the case of liquids, gases and plasma we cannot predict their motions precisely, but only probabilistically, with a range of outcomes, meaning that we can only say how they are likely to move and their motion is not to a single position but is a wave.

Unlike the Cartesian diagram that is intended to represent the relation of 2 independent variables, the diagram above and all the following diagrams are intended to represent the relations of 4 independent variables, with each variable on its own axis and with vectors going in both directions on each axis to represent positive and negative values.

So the above graph can be used as **a "dimensional" graph, i.e. in terms of showing how many directions and dimensions are involved in the motions of matter synchronously.**

2. The Unified Theory of All

Physics: There are 4 fundamental forces in the Universe: Gravity, Electromagnetism, the Weak nuclear force and the Strong nuclear force.

-Only gravity and Electromagnetism are long range forces and we know that they are in opposition— they do opposing functions— as indicated by the opposing signs in the equations of the inverse of distance squared laws of Newton for gravity vs. for electromagnetism of Coulomb. The opposition in the signs is due to the fact that similar charges in electromagnetism repel rather than attract each other as is the case for masses. That opposition in these two equations was a key reason why Einstein looked for a better theory that may unify these two forces, yet despite having discovered General Relativity for gravity, he could not find a Theory to unify Gravity with electromagnetism and the other 2 (quantum) forces. Gravity pulls down; we are able to stand and lift things because of the electromagnetic force.

-The opposition in the function of the short range "within" the atom forces is also clear: The Weak force by causing spin causes emission of radiation and so reduces by a certain half-life the mass of the nucleus, is in opposition to the strong force that holds (the mass of) the nucleus together.

The reason the Strong force was placed on the right, (rather than left) has no validated evidence, yet, but the hypothesis is that because its local internal 3 symmetry it operates more like gases rather than liquids to which the Weak force seems correlated in terms of directions of synchronous motion involved, as follows:

3. Physics

There are only 4 known and fully proven fundamental forces in the Universe: Gravity, Weak, Strong and Electromagnetism.

It is these 4 Fundamental forces that cause motion and transform energy. Gravity is described by Newton's laws and by General Relativity and is deterministic and is thus very distinct, and has not been unified with the 3 other forces i.e. Electromagnetism, the Weak and the Strong force, that are probabilistic Quantum Forces and are described by The Standard Model.
Do the Forces correspond to what Nature shows?

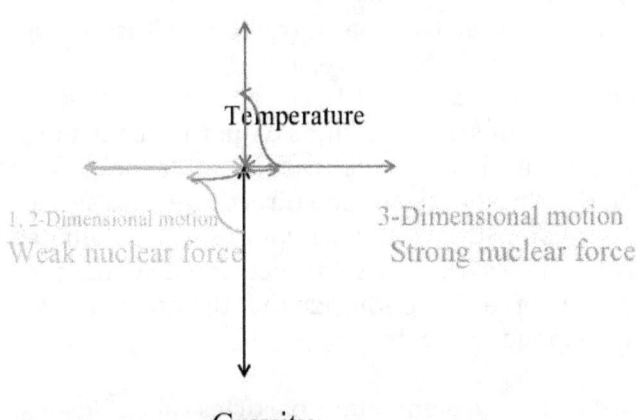

1; (and 2) and 3 Dimensional motion
Electromagnetism

Temperature

1, 2-Dimensional motion   3-Dimensional motion
Weak nuclear force   Strong nuclear force

Gravity
1-Directional motion

Elementary particles

There are 12 elementary particles called fermions:
6 leptons and 6 quarks.

The 6 leptons are:
3 leptons with charge -1 of different sizes that are: Electron,
(e); Muon, (μ); Tau, (τ), that for sure are operated by
electromagnetism. Their sizes and corresponding masses are
not accurately represented on the graph because they have
large ratios with the Muon being about 200 times heavier
than the electron and the Tau about 15 times heavier than
the muon.
So, the graph below must be considered logarithmic.
3 other corresponding leptons called neutrinos v_e, v_μ, v_τ that
respond only to the (nuclear) Weak force.

And there are 6 quarks;
3 up quarks: up (u); charm (c); top (t); that are operated by
the (nuclear) Strong force, and
3 down quarks: down (d), strange (s), and bottom (b).
Combinations of three kinds of quarks make things like the
proton (uud) and the neutron (udd.) It should be noted that
quarks have not been seen directly yet and so the "quark
theory of matter" is still a hypothesis. It is founded on the
mathematics i.e. on the 3 Dimensional Symmetry of the
Strong Force. I hypothesize that the down quarks
correspond to gravity.

There are also elementary particles called Bosons. The
4 <u>fundamental forces</u> of nature are mediated by <u>gauge
bosons</u>, and mass is believed to be created by the <u>Higgs Field</u>.
Consistent with the <u>Standard Model</u> the elementary bosons
are shown below in caps and bold.

All these elementary particles and their relationships are
shown on the graph below:

Elementary particles of matter

There are two classes of elementary particles: Fermions and Bosons. There are 12 fermion elementary particles, according to the Standard Model, shown on the Cross below: 6 leptons; and 6 quarks.

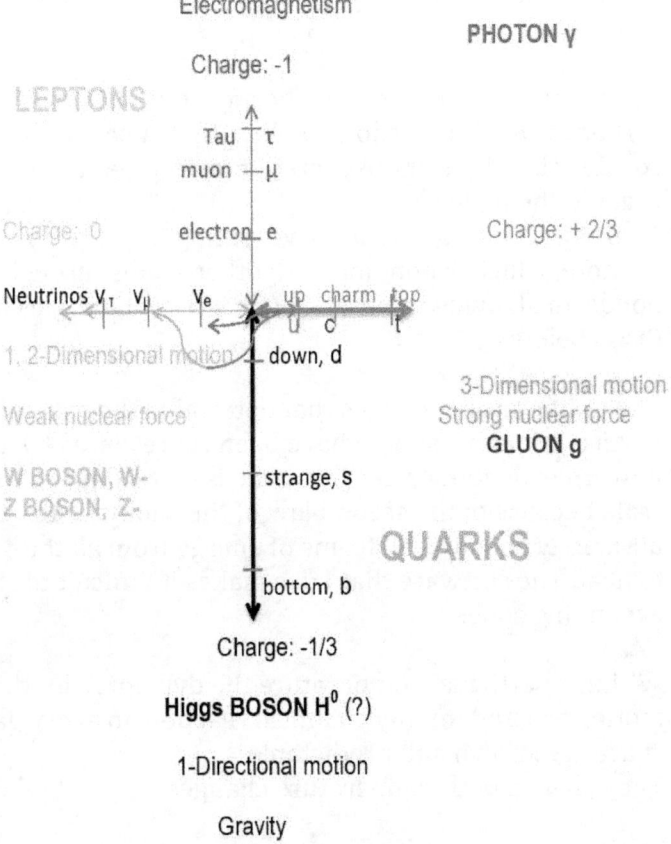

1; (and 2) and 3 Dimensional motion

Electromagnetism

PHOTON γ

Charge: -1

LEPTONS

Tau τ
muon μ

Charge: 0   electron e   Charge: + 2/3

Neutrinos Vᵧ Vμ   Vₑ   up charm top
                            u   c   t
1, 2-Dimensional motion   down, d

                          3-Dimensional motion
Weak nuclear force   Strong nuclear force
                          GLUON g

W BOSON, W-   strange, s
Z BOSON, Z-

                          QUARKS

              bottom, b

Charge: -1/3

Higgs BOSON H⁰ (?)

1-Directional motion

Gravity

We cannot show all 4 dimensions— the 3 of space and time — fully, on a flat 2 Dimensional screen. To simplify, we represent the 4 Dimensions on 4 axes, as if flat on a 2 Dimensional screen.

Time, has an arrow, i.e. goes in one direction only, and curves space around mass.

The arrow of time is caused by the second law of thermodynamics that over time converts all energy to heat. Every part of space is connected through electrons with other parts, by bonds of opposing charges that are correctly described as a Cross that includes sub-crosses at every point on the Cross.

The atom at the center could be any atom; however the carbon atom, C, as is shown below, that is balanced with 4 covalent bonds, is at the center of (and necessary for) all organic chemistry.
Therefore, everywhere and everything in our body there is an atom, which is bonding with other atoms through several bonds. In all living things, through 4 basic bonds, as in the Cross below.

The temperature changes that determine the state of a particular matter should have been represented by an upward expanding cone. The cone is expanding on the grand scale because of the second law of thermodynamics that ultimately converts all forms of energy from all the 4 forces to heat. The software that I use makes it difficult to show the expanding cone.

Within a particular temperature the dynamics, i.e. the motions, of and for any chemical reaction, in every part of space are known and predictable.
They change with temperature changes.

4. Chemistry

Electro-magnetism is also called the electrochemical force because it operates all chemical reactions, also.

The central component determining chemical reactions are the 4 quantum numbers, determining the motion and pathway of any electron around any atom.

So, considering the nucleus at the intersection, the 4 quantum numbers that determine the motions of the electrons can be described on the Cross:

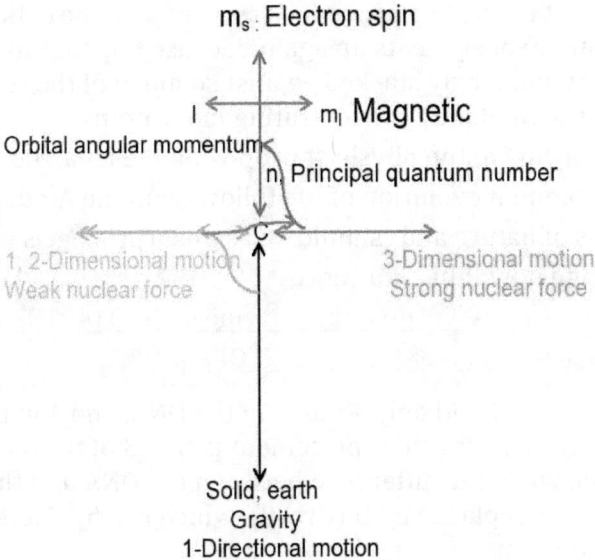

1; (and 2) and 3 Dimensional motion
Electromagnetism

m_s : Electron spin

l
Orbital angular momentum

m_l Magnetic

n Principal quantum number

C

1, 2-Dimensional motion
Weak nuclear force

3-Dimensional motion
Strong nuclear force

Solid, earth
Gravity
1-Directional motion

"From the standpoint of physics, there is one essential difference between living things and inanimate clumps of carbon atoms: The former tend to be much better at capturing energy from their environment and dissipating that energy as heat.

We can show very simply from the formula that the more likely evolutionary outcomes are going to be the ones that absorbed and dissipated more energy from the environment's external drives on the way to getting there," J. England, at MIT, said.

"At the heart of Jeremy England's idea is the second law of thermodynamics, also known as the law of increasing entropy or the "arrow of time." There are more ways for energy to be spread out than for it to be concentrated. Thus, as particles in a system move around and interact, they will, through sheer chance, tend to adopt configurations in which the energy is spread out. Eventually, the system arrives at a state of maximum entropy called "thermodynamic equilibrium," in which energy is uniformly distributed. A cup of coffee and the room it sits in become the same temperature, for example. As long as the cup and the room are left alone, this process is irreversible. The coffee never spontaneously heats up again because the odds are overwhelmingly stacked against so much of the room's energy randomly concentrating in its atoms.

According to the physicist proposing the idea, the origin and subsequent evolution of life follow from the fundamental laws of nature and "should be as unsurprising as rocks rolling downhill." For more

see: http://www.quantamagazine.org/20140122-a-new-physics-theory-of-life/#ixzz3LOkZqXWg

There are 4 and only 4 bases in the DNA, and 4 in the RNA. The rational for their placement is that 3 of the bases are the same while the difference between the DNA and the RNA is that T is replaced by U (Urasil), which is why T it is placed at the bottom.

The hypothesis about their placement needs to be tested.

5. Biology

The four bases of the DNA that records all the history of life, as life goes on, and precisely identifies each human and animal are: adenine (abbreviated A), cytosine (C), guanine (G) and thymine (T). They bond as A-T; G-C.

As temperatures change there is a helix, up and down the Cross, and up and down the temperature cone, a double-helix, that is formed by the bonding of the 4 bases, which, again, because of limitations of my software I showed by the curved arrows.

The 4 bases of the DNA and how they bond, with only one side of the helix to avoid clutter, can all be represented on the Cross, as follows:

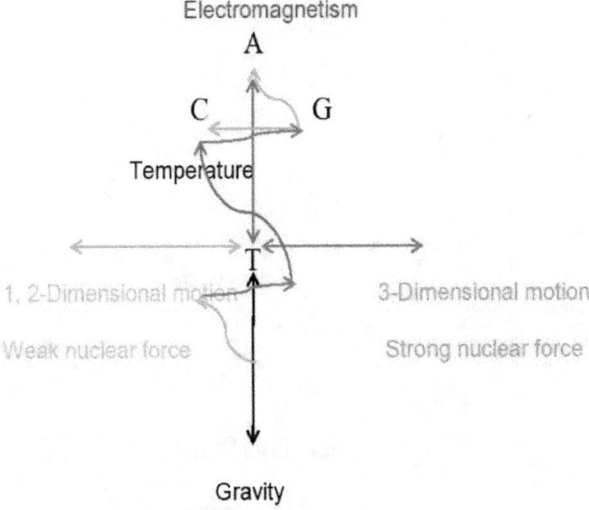

1; (and 2) and 3 Dimensional motion

Electromagnetism

A

C   G

Temperature

1, 2-Dimensional motion   3-Dimensional motion

Weak nuclear force   Strong nuclear force

Gravity

1-Directional motion, time

6. Psychology

The Cross has the individual at its center, showing the forces on and the functions of each individual.

There are 4 corresponding functions of humans, according to C. Jung: the functions of perception of sensing (S) vs. intuiting (I); and the functions of judging of feeling (F) vs. thinking, (T).

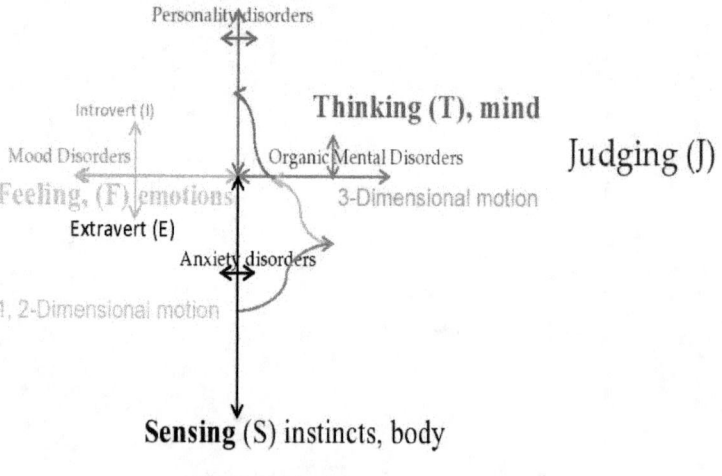

Perceiving (P) (the 2 vertical axes i.e. intuiting and/vs. sensing)

Intuiting, (I) intent, conscience, character, soul

1, (and 2) and 3 Dimensional motion

Personality disorders

Introvert (I)

Thinking (T), mind

Mood Disorders   Organic Mental Disorders   Judging (J)

Feeling, (F) emotions   3-Dimensional motion

Extravert (E)

Anxiety disorders

1, 2-Dimensional motion

Sensing (S) instincts, body

1-Directional motion

These along with two additional dichotomies— i.e. bipolar dimensions in which each pole represents a different preference— the preference in attitude of Extravert (E) vs. Introvert (I) and the preference of Judging (J) vs. Perceiving (P) that was added by Briggs Myers, form the typology of the 16 "personality types" that are commonly used in Psychology and in business.

To understand life one must know your-self. Once you "know your-self," as the Mystics inscribed for the Oracles of Delphi, you know the internal opposition of the duality "within" any indivi-dual. It's a duality such as that of your left and/vs. your right side and a duality such as that of light being both a particle and a wave.

That internal opposition along with an opposition with the outside world forms the 4 axes, the 4 dimensions, of the spiritual Cross of each one of us.
Aren't intuition, conscience and intent a function of the soul; feelings a function of emotions i.e. "the heart;" thinking a function of the "mind"; and instincts and sensing a function of the body?
Does intuition work like fire? Like a light-bulb lighting up? Are emotions like fluids, moving as surfaces of blood through the heart?
Do thoughts occupy all available space, like the air moving in all 3 space dimensions (as sound waves do) at the same time? That's the hypothesis of their correlations.
Emotions are expressions of one's truth; thoughts express reason; and intent expresses one's love or hate.
It's these two oppositions that cause friction, stress and conflicts.

However, their causal relationship is that they are derivatives of each other.
In the form of integrals: The truth of truths is reason. The reason for reason is right reason; the reason for right reason is love.

Psychiatry

There are large numbers of psychiatric disorders described in the DSMR (Diagnostic Statistical Manual by the American Psychiatric Association) that cover the whole Cross. To avoid clutter only 4 common categories of: Anxiety, Mood, Organic Mental and the Personality Disorders, are shown, with their obvious correspondences and correlations.

The narcissistic, histrionic, and borderline schizophrenic personality disorders that are mentioned in the text are a sub-cross of the Personality i.e. the character disorders, that are disorders of "moral content" and are difficult to treat.

The advantage of showing all the mental disorders on a 4 axis graph is that each factor or function or preference can be quantified, shown as an independent vector (as shown in the investing graph,) and the impact of all the vectors can be summarized quantitatively in one graph.
It also shows the correlations and causation with all other disciplines and resulting vectors.

Mathematics

Each of the 4 axes of a Cross, depending on the discipline or subject that is examined should describe an independent variable as a vector, in either direction to allow for both positive and negative numbers.

The addition of the 4 independent vectors on 4 dimensions can be complex.
The mathematical solution of functions that describe and provide equations of motions of dynamical systems with 4 independent variables (x, y, z and t) is by what is called a Lagrangian. Use as t, the vector of the bottom axis.

In Quantum Theory when more than two orthogonal vectors are to be added the cosine square law is used.

Each of these 4 axes is orthogonal to the other as Special Relatively shows about the 3 space dimensions and time. For example, the top axis is not actually in the same line as the bottom but is orthogonal to it.

Yet, in my estimation just adding the vertical axes as if in one dimension and the horizontal axes as if in a single dimension and then adding the two vectors geometrically produces adequate, first approximation, results, which for me being a lazy Greek are close enough.

However, for more accurate answers computerized solutions based on Lagrangians can do the calculations
and display the sum of the 4 vectors.

Note

The 4 factors— independent variables, forces, vectors— constituting each Cross of each Academic discipline are well known, proven, public knowledge that can be easily checked from the Bibliography or on the internet and do not need to be cited individually.

The correct or not placement of each of the 4 forces, factors to each of the 4 axes and the resulting correlations have not been fully proven yet. I hope that they get tested independently to be confirmed and proven or disproven.

7. Interpretation of symbols in the Bible

What do the four horsemen of the Apocalypse represent? White is the combination of all the colors of light in high speed. The rest: black i.e. African, red Native American, yellow Asian-Chinese are self-explanatory. They do represent the four races of humanity.

What do the colors of Babylon or of the red Dragon represent?

What does the Beast from the sea mean?

What does the Beast from the earth mean?

What does Baptism by water and/or by fire mean?

The graph shows it.

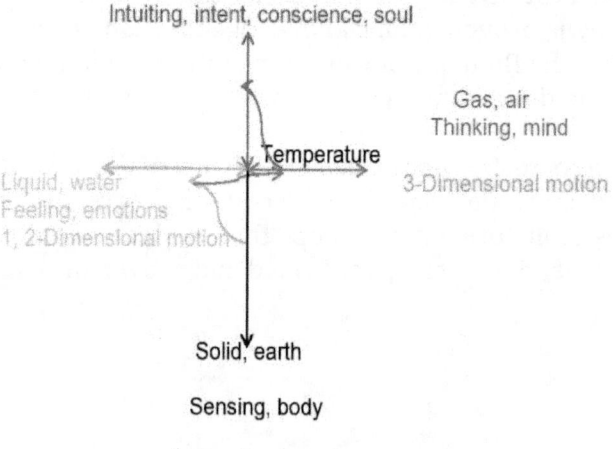

1, (and 2) and 3 Dimensional motion

Plasma, fire

Intuiting, intent, conscience, soul

Gas, air
Thinking, mind

Temperature

Liquid, water
Feeling, emotions
1, 2-Dimensional motion

3-Dimensional motion

Solid, earth

Sensing, body

1-Directional motion

One may note that Nature describes things by black and by the colors of light, that are derived by mixing the 3 primary colors of yellow, which is not shown well because it doesn't appear on the graph adequately that along with green forms the left axis; of red on the top axis; and of blue on the right axis, as shown above.

Even though there are hypotheses about each primary color having a different emotional impact I have not found scientifically validated data on it. The following seem reasonable hypotheses:
-Red (on the top axis) is the color of fire and blood, so it is associated with energy, war, danger, strength, power, determination as well as passion, desire, and love.

-Yellow (on the left axis) is the color of sunshine. It's associated with joy, happiness, intellect, and energy. Green is the color of nature. It symbolizes growth, harmony, freshness, and fertility. Green has strong emotional correspondence with safety.

- Blue (on the right axis) is the color of the sky and sea. It is often associated with depth and stability. It symbolizes trust, loyalty, wisdom, confidence, intelligence, faith, truth, and heaven.

This Cross is manifested in every aspect of Creation, Nature, academic discipline and of life, and unlike anything else, shows the causation, connections and correlations between **physics, chemistry, biology, psychology, business, economics and politics** that until now could not be explained, which is why these disciplines had been considered "completely different" disciplines. Yet, as you can see, they each and all operate through the same Cross.

In simple language what the Cross describes is that everything and everyone, in whatever context one may choose to look at, is being pulled and/or pushed at any time in 4 opposing directions and dimensions, spiritually but also physically by the 4 Universal forces, through the 3 dimensions of space and by time.

This is not just the only provable Unified Theory of Physics but to the best of my knowledge it is also the only Unified Theory of all major academic disciplines.

8. Business

Is there a correspondence of how the 4 States of matter function to how businesses function? Revenue growth requires reduction in cash flow; and Income growth requires reduction i.e. investment of equity.

Revenue growth

Fire, Intuiting

1, (and 2) and 3 Dimensional motion

Air; Thinking

Equity ← → Income growth, EPS growth

Water; Feeling

1. 2-Dimensional motion   3-Dimensional motion

Solid, earth, Sensing

1-Directional motion

Cash flow change

Book value of equity; a multiple of annual cash flow; a multiple based on revenue growth; and a multiple (P/E) of EPS are the four basic ways of estimating the Discounted future Cash flows and thus the value of a company. Which of the methods is more accurate in each particular circumstance depends on which of the four stages a company is in.

For start-ups and companies with little income, their company valuation is based on book value of their equity; for the high growth stage company value is by revenue growth; for the mature stage the value is by a multiple of cash flow; and for the "old" stage by the P/E (Price to Earnings ratio.) All valuation methods are taken into consideration each time with emphasis on the one that corresponds to the stage of the company.

As with every one of the cases above i.e. the strategies, the states of matter, the human functions and business evaluation graphs the axes that are in opposition to each other represent the direct opposition in what they describe. While at any point of time there is a double-opposition, a Cross, because "everything is in motion," as Heraclitus explained, the Cross also represents the 4 phases of any lifecycle, with each axis representing a different phase.

So that if left without intervention, the actual motions are "up" on the Cross; with the reaction "down" the Cross, like a double-helix. (Like the "ladder" to and from the heavens that Jacob dreamed.)

Because there can be both positive and negative numbers on each axis, by convention we use: up- positive, down-negative; left- negative, right-positive. Therefore the cross is set for the "best balances" to be in the upper right hand.

When people say "our strategic direction" they talk about it but they cannot show any actual geometric direction graphically.
This Cross allows one to actually show a precise direction, with a precise angle and a precise vector size for each strategy.

The most significant cause of failure of organizations is the misalignment of the objectives with the strategy or with the organizational structure or with the implementation tactics i.e., the action plans. Therefore a particular strategy requires a particular emphasis, focus, on a particular function and a corresponding increased funding to that function in the organization to get alignment as shown on the next Cross.

9. Business strategies

There are only 4 (corresponding?) basic company strategies: Increase Market share vs. increase income (profit); and Cash growth vs. Revenue growth, as follows:

Revenue growth

Intuiting

1, (and 2) and 3 Dimensional motion

Engineering, Design

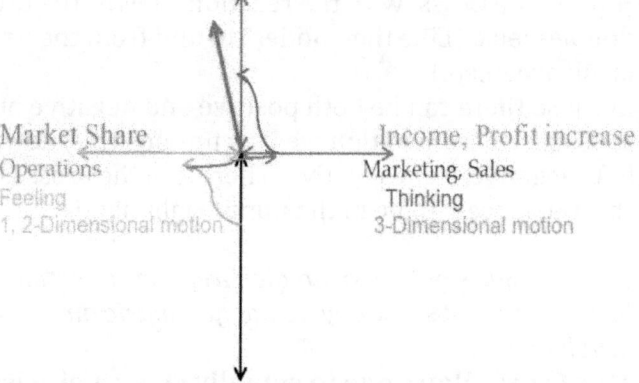

Market Share   Income, Profit increase
Operations   Marketing, Sales
Feeling   Thinking
1, 2-Dimensional motion   3-Dimensional motion

Finance, Administration

Cash flow change

Sensing

1-Directional motion

This allows precise graphic description of a strategic direction, as the above.

10.   Economics

The 4 basic factors determining the state of an economy are:
Unemployment vs. inflation, (excessive money growth) that are managed by
monetary policy through changing interest rates;
and GDP growth vs. wealth distribution, that are managed by fiscal policy.
Can both high GDP growth and high wealth distribution be accomplished?
Yes, but only if one knows what they do.

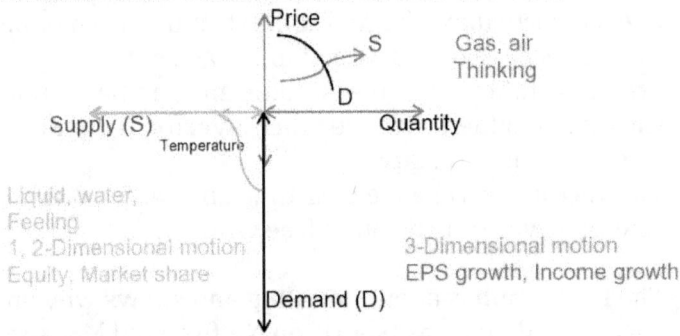

1, (and 2) and 3 Dimensional motion
State of matter   Plasma, fire

Individual: Intuiting

Business: Revenue growth

Economic factor: Inflation minus interest rates

Price

S

Gas, air
Thinking

D

Supply (S)   Quantity
        Temperature

Liquid, water,
Feeling
1, 2-Dimensional motion   3-Dimensional motion
Equity, Market share   EPS growth, Income growth

Demand (D)

Wealth Distribution   GDP growth

State of matter: Solid, earth
Human function: Sensing
Dimensions of
synchronous motion: 1-Directional motion
Business factor: Cash flow

Economic factor: Unemployment

It is the superimposition of the Demand curve from the bottom axis relative to Price-Quantity, and of the Supply curve from the left axis (again relative to the Price-Quantity,) to the upper right quadrant that makes the usual economic graphs.

So, this is a way to in one graph see all the 4 central factors that are in opposition with each other, such as in a Cross, in all the central aspects of the Universe and of life from the Thermodynamics of Nature, to Physics, to the Human constitution and functions, to business evaluations and strategies and to economic evaluations.

Each Cross shows the 4 primary opposing driving forces in the context of each academic discipline and the correspondences and correlations in terms of the top 3 Axes; and the causation in the case of the bottom axis, to Crosses of other disciplines.
In each particular case and context, and at each point in time the intensity, size and direction of each of the 4 opposing vectors is different, is measurable and can be graphically shown and added with the other 3 vectors to get to an overall direction and intensity.
The vector sizes change over time and can be rationally changed over time by one's free will.

The next graph is about investing and shows why on particular date I was "short" on Netflix; sold MSI and held for a buying opportunity for CMI. The software program that I developed and use connects each "company Cross" with a "larger" Cross by industry, also.

As of this writing, I was wrong in being short on Netflix. Relative P/E is a measure of the confidence the market places on a CEO/company.

11.   Investing

Investing must take into account the economic fundamentals; industry trends and the drive of the CEO. The value of a company is best estimated through discounting future cash flows which is not practically feasible unless one is an insider. As an outsider, there are 4 opposing factors determining the value of a company. Revenue growth vs. the Cash flow from operations; because higher growth rates demand more cash investments and reduce cash flows from operations. On the other axis there is Earnings per share (EPS) growth vs. the P/E that one has to pay for such earnings growth.

Evaluating the opposing factors for a buy/sell decision: Example: As of 11/ 1/ 2014, IBM vs. GE vs. P&G vs. Netflix vs. CMI vs. MSI.

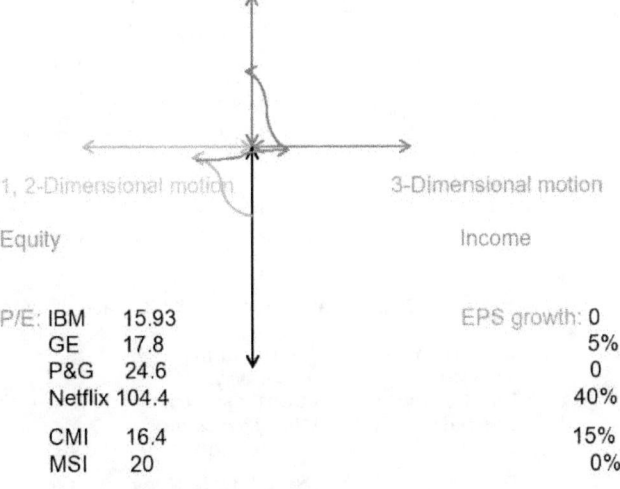

Revenue growth: 0, 0, 1%, 25%, 12%, 0%,

1, (and 2) and 3 Dimensional motion

1, 2-Dimensional motion   3-Dimensional motion

Equity   Income

P/E:	IBM	15.93		EPS growth:	0
	GE	17.8			5%
	P&G	24.6			0
	Netflix	104.4			40%
	CMI	16.4			15%
	MSI	20			0%

1-Directional motion, sensing

(Times) X Cash Flow from operations; 11,13,17, 440, 13, 27

12.   Politics

The four basic separate and/or conflicted groups are the "Left" vs. the "Right" and/vs. the Independents and/vs. the non-voters, which can be measured at any time and territory and placed on the Cross.

Everything in one Graph
Space Directions (and dimensions) of Synchronous Motion Involved

1, (and 2) and 3 Dimensional motion

State of matter: Plasma, fire
Physics force: Electromagnetism
Individual: intuiting-soul
Business: Revenue growth (at a particular inflation)
Economics: Inflation

Politics: Independents

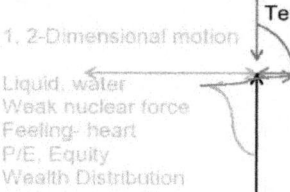

Temperature

1, 2-Dimensional motion   3-Dimensional motion

Liquid, water   Gas, air
Weak nuclear force   Strong nuclear force
Feeling- heart   thinking- mind
P/E, Equity   EPS growth, Income growth
Wealth Distribution   GDP growth

Liberal "Left"   Conservative "Right"

1-Directional motion of all its 3 space dimensions

State of matter: Solid, earth
Physics: Gravity
Psychology: Sensing- body
Business: Cash flow change
Economics: Unemployment

Politics: Not voting in protest or not interested.

The often neglected are the non-voters which for example in the 2014 elections in the US were over 60% of the population.

Some of the academic disciplines were not included in this last Exhibit to avoid clutter but once one understands the correlations, one may apply them as a Cross to any discipline, field of knowledge, circumstance, space and time, person and/or organization, at any level of detail.

Philosophy

The above correlations shown on the Crosses are from the main fields of knowledge of Physics, Chemistry, Biology, Psychology, Business, Economics and Politics are in terms of ways, of dynamics of motion and correlations and are not causation, except in the case of the bottom axis, in which case there is a cause-effect. The top three axes are correlations.

"Everything comes from or through its opposite," said Socrates.
Everything exists within opposites; and each is seen in the context of its opposite.
A simple example of this from Nature is that at night one can see the lights, while during the day (in light) one can see the dark things.

In the Universe those oppositions are between States of the same matter.
There is a polarization within each nucleus not only among opposing charges but also by opposing forces, i.e. of the short-range Weak and Strong nuclear forces. There is also another polarization between the nucleus and its electrons within each atom not only in terms of opposing charges but also by the opposing long range forces of electromagnetism and of gravity.
It is these two separate polarizations that the Cross, as above, most simply and accurately describes.

As you might know Special Relativity is an extension of the Pythagorean into 4 dimensions. The Mystic Pythagoras also taught about Oneness, correctly claimed that the earth is round and explained the mathematics and geometries of music. He first used this taxonomy, (categorization, organization) on a 4 axes Cross, based on the 4 states of matter.

The equal of all 4 Axis Cross was also the symbol of Zeus, as evidenced by a marble cross found in Crete, for Zeus, (currently in the Heraklion Museum) dated to 1600 B.C. during the Minoan civilization.

The Cross is also the symbol of all Christianity.

Because of what happens "within" each atom (and for the same reason) there's also a corresponding internal and an external polarization within and among atoms, and molecules, and within and among things, and "within" each person, ("atomo" is the Greek word for a person) and within and among people and within and among organizations.

The correspondences, which are the causation in terms of the bottom axis of solids; and are correlations among the 3 top probabilistic quantum axes of light have been theorized in these Exhibits but have not been fully proven yet.

What has been proven is that any issue, any academic discipline, any person or organization can be best described on a Cross; at any level of detail with Sub-crosses; in any context of a "larger" Cross.

In the context of the 4 states of matter, solids operate in a linear, deterministic, evolutionary manner; while the three other states and particularly, plasma, fire, in a probabilistic, non-linear, revolutionary, quantum, "leap" manner, causing big, sudden changes.

In the context of the 4 fundamental forces of physics, gravity that is dominant in operating the solids also causes linear deterministic continuous evolutionary motions; while the other 3 forces operate non-linearly, are many orders of magnitude faster, are clouds with wave and cloud like multi directional-dimensional probabilistic motions, are quantum, that means discontinuous, "leap" and so are revolutionary.

The Unified Theory of All; The Unified Theory of Life is: Revolution and Evolution.

It is proven by history and by each of the 12 Crosses describing the driving forces, factors, of each of these 12 Academic Disciplines.

The Spirit and State of one, determines one's interpretations. If you disagree with me, your opposition proves the Cross. There can only be unity about the cross. Only by unity about our cross we're free from the cross.

As will be shown subsequently there is a right and Just Logic of reciprocity that governs the Universe.

Right (for the other) **intent expressed through right thinking, right judgments by the right criteria followed by right action for and in the right context is self-evidently the only right philosophy that can produce the right results.**

The right intent is love of others.
The right results are the change of the current spiritual reality into being the Eternal Truths –shown later— by which humans operate and the resulting changes of the physical reality so that there is harmony for all.
The right context, are the Independent Demilitarized Disarmed Holy States.

To have real life practical meaning this right philosophy needs to define and know (and it does) Right Logic and Righteousness.

Right Logic and Righteousness is the Truth of Truths; Christ Jesus.

Religion

The philosophical proof that **love, right and true** are each necessary, are the only ones necessary and must be unified for a complete and cohesive, right philosophy is quite self-evident but I will present it.
- Anyone or two of these three aspects on their own are either not loving or wrong or false and thus are inadequate and incomplete. For example loving rightly is not too valuable if it is not true. Or truly loving isn't too valuable either if it is wrong, such as truly loving doing bad. One can make similar statements for all combinations.
  This proves that each of love, right and true are necessary, for any to have its right meaning.
- Given that The Truth includes all the Eternal and all temporal truths, i.e. facts, these three aspects of the Holy Trinity are the only ones needed.
- Any communication that isn't true or isn't right or doesn't come from love defeats its cause, means and/or purpose.
- These three Aspects of the Holy Trinity are derivatives of each other and are respectively the cause of any communication, the means and its result (objective, purpose.)

"God is Love." (1 John 4:8)

Love is the intent, cause and purpose; Right is the reasoning, the means; and the existence of Love in Truth, both spiritual and physical, is the result, the end, of the Expressions of God.

"From the bottom up:" Truth is the cause of any communication, right is the means and love is the result. This proves that all these three Aspects of the Holy Trinity need to be in unity for any of God's Aspects to be.

In other words, the evidence that right is right is the truth.
The evidence that the truth is true is right reasoning.
The evidence of Love is righteousness.
The relationship of loving intent, through right thinking producing the right spiritual and physical truth, is the aligned oneness of spirit and body.
Philosophy means the love of Wisdom.
The love of Wisdom is inevitably bound to find and found the Wisdom of Love.
The Wisdom of Love, the Holy Wisdom, Agia Sophia, is Righteousness.
True philosophers, who have logic as their only tool, to be right were bound to find and submit to righteousness, the Christ Jesus.

Everything, every discipline and everyone has a spiritual Cross, as above, being pulled and pushed in opposing directions all the time.
"Within" each, is one polarization; and vs. the environment, is the other polarization resulting in oppositions, contradictions, tensions, difficult decisions and conflict.

The physical Cross was a Roman invention and was the death sentence imposed on non-Roman criminals and it is modeled, with two cross bars, after the Cross like physical bone, skeletal structure of humans.

Assuming that what Nature shows with how matter, in each of its 4 States moves differently and distinctly, is not random then there must be correlations.
The suggested correlations are with how "matter" in every part of the Universe and of existence has 4 corresponding States and understanding those 4 States that produce oppositions in life, is critical to a self-examined life, which is the only life worth living, per Plato.

All Crosses are Sub-crosses of a single Cross, with 3 sub-divisions on each Axis of the Cross, just like the Cross of the

12 fermion elementary particles, each cross-point of which is in it-self a cross, which is in turn sub-divided as above and so forth, making the net of space over history. This pattern within a pattern is characteristic of chaotic phenomena as the theory of Chaos explains.

Depending on one's own state of being each, (naturally,) perceives the Crosses and reality, differently.

Finding and correcting the biggest imbalances to equalize the 4 axes of the cross is how its burden is minimized.

Because the biggest imbalance is usually that the bottom axis of the solid material dominates the usual correction to balance one's cross is reduced emphasis on the material and increased emphasis on the intellectual, emotional and moral.

Life is not about getting to a static balance, like a statue but rather it is about causing and/or getting through highly imbalanced situations in a dynamic, moving, balance. **Life works through balanced imbalances.**

One of the applications of understanding the correlations among the crosses is to understand that a change in someone's thoughts and/or emotions does not just affect the state of balance and direction of that person but that it has reverberations into all other crosses, (that appear at the same time as if by coincidence) which can be understood and predicted by understanding the correlations above.

Also, if one does not understand any particular factor on any Cross they may understand how it works by understanding its corresponding factor on any other Cross.

Another application of the Cross is that it is a way to understand the nature of the cultural and social differences between West and East and between North and South in global context, within each continent and within each nation. In the context of geography one may place N-S; W-E on the Cross and by observing the corresponding correlations on the other Crosses one may get an understanding of the nature of the natural social tendencies that polarize North with South, and West with East.

How can we make sense of everything? Is there any other theory showing the findings and correlations among all these fields of knowledge?
I come back to this graph. If there is such a thing as a **Golden Compass** these corresponding Crosses are it.
There is a lot of work that needs to be done in testing and proving the above causation in the case of the bottom axis and correlations for the other 3 quantum axes; but they must be done by others because I have gone as far as I can. If the above is not correct then what is?

What are the correlations among the academic disciplines or are all the teachings from our education incoherent, self-contradicted, unrelated and/or random and therefore not understandable?

Because each human has to use knowledge from each of these academic fields to live in this day and age if the sum of what people are taught is broken up incoherence they will end up broken and incoherent.
If incoherence is the case among the academic disciplines, then the most famous of your educators must be not the ones that really know most but the ones who best pretend to know.
Is everything we know consistent with the Cross of Christ or not? It is. The evidence has been shown from almost all fields of knowledge.
This is the Unified Theory of all; of everything; of all fields; all people; of all life that is consistent with the Cross of Christ.
If the teaching is not consistent with the Cross of Christ it is wrong because only the Christ Jesus is Absolutely Right. This will be proven, subsequently.

The evidence of God is everywhere physically and is available to each spiritually. However, if a non-believer doesn't accept any of them then the evidence of the existence of God is the real life and afterlife of Jesus.

(Now that you have seen how big the Cross of this whole world is, you must be feeling at least some sympathy for poor Atlas having to carry the burden of it!)

"Whoever wants to be my disciple should deny himself, *lift up* his *cross and follow me.*" (Mark 8:34 and also in Luke 9:23)

Understanding the opposing factors and the Cross of each at the personal, relationship, business and political level has the purpose of lifting one's Cross.

Once the 4 axes are symmetrical it is not difficult by "spinning" them correctly to use the 4 axes as propellers, (helix is the Greek word) and creating an upward spiral in which every action has the added effect of accelerating positive action in at least one of the other axes.

Creating an upward spiral, a virtues cycle, is done by a coherent strategy in which the objectives are aligned straightly with the incentives-disincentives and the everyday actions and by correct phase timing.

That is how to "lift your Cross."

To be in oneness one must be in oneness within themselves and in oneness with their environment also.

Lifting one's Cross has the purpose of going somewhere. Where to go is to follow spiritually The Christ Jesus, your conscience and your right reasoning honestly into the Heavenly Kingdom of Eternal Truths and to physically follow Him into one small physical Independent Demilitarized Disarmed Democratic State, within the territory of your nation.

This is central and the only way for people to get to oneness and to eternal life in joy and peace.

3. Motives; Intent: What do you want?

Do most women prefer rich men or poor?
Do you like people who are "happy" and entertaining or
those who are unhappy and depressed?
Are most women attracted to strong or to weak men?
These questions are somewhat rhetorical because we have
seen many models, actresses, singers, groupies or other
women chase after the rich, the happy, famous and the
strong, even if they look unattractive, yet I have yet to see
women chase after any poor, any depressed or any weak to
love or to have sex.

Did Jesus say "blessed are the rich, the happy and the
strong?" or did he bless the poor, the unhappy and the weak?
Luke 6:20-23 says:
"Blessed are you who are poor,
    for yours is the kingdom of God.
Blessed are you who hunger now,
    for you will be satisfied.
Blessed are you who weep now,
    for you will laugh.
Blessed are you when people hate you,
    when they exclude you and insult you
    and reject your name as evil,
        because of the Son of Man.
Rejoice in that day, and leap for joy, for behold, your reward
is great in heaven; for so their fathers did to the prophets.
A very similar passage is in Matthew 5:3-5:
"Blessed are the poor in spirit, for theirs is the kingdom of
heaven. Blessed are those who mourn, for they shall be
comforted. Blessed are the meek, for they shall inherit the
earth. Blessed are those who hunger and thirst for
righteousness, for they shall be satisfied."

And then Luke 6: 24-26
"But woe to you who are rich,
    for you have already received your comfort.

Woe to you who are well fed now,
   for you will go hungry.
Woe to you who laugh now,
   for you will mourn and weep.
Woe to you when everyone speaks well of you,
   for that is how their ancestors treated the false prophets."
Why did Jesus bless the poor, the unhappy and the weak? Is
Jesus right or is he wrong?

Who are the "strong"? The physically strong are, for example,
the sports athletes who can physically beat others. The
legally strong are those who have positions of authority over
others as do the politicians, who by having authority over the
physical force of law enforcement agencies are much
stronger than the physically strong.
The strong, exercise their strength by lording over, by being
the bosses of, and by controlling the weak. The strong, such
as the politicians, make up, and through the police enforce,
their laws by which the weak— and not really Congress or
the President— have to abide.

Are those strong who are higher up in the political hierarchy,
the rich in the economic, and the happy in the social-
entertainment hierarchies the ones that are more likely or
less likely to be more hypocritical, mean, beastly, deceitful,
selfish, double talking and double dealing?
Are the rich, the "happy" and the strong the ones that are the
abusers, the deceivers and the ones that hurt others or is the
problem the poor, weak and depressed which is what the
perpetrators keep saying?
Is the problem that the poor, unhappy and weak are so
because they are lazy, uneducated and have "issues" as the
rich, happy and strong claim?
Or is the problem the hypocrisy and double talk about "being
created equal" but "we govern" with the implication that "we
are better" and the abusiveness, financial abuse, violence,
hurtfulness and demeaning of others by the rich, "happy" and
strong?

If one is to prefer some over others then one must examine whether one is choosing and judging by the right criteria and not by the wrong criteria.

If material short term survival is one's criterion, then the rich, the strong and happy are the better choice but they are the worst choice if the criterion is what is morally right, which is central to long term survival.

If a woman's criteria for choosing and having sex with someone are her instincts and what appears better, then it is natural to be attracted to the rich, happy and strong.

But then she has to deal with the consequences of getting over-controlled, mistreated, deceived, abused, hurt and treated like a material object.

If one's criteria regarding choosing a friend or husband are goodness, selflessness, being morally right and one's spiritual well-being then choosing the "good" guys who (as the saying goes) "come last" in this world, the weak, the poor and the depressed, is the right choice.

In my judgment, which coincides with Jesus' statements, our political and socioeconomic leaders, the strong, rich and happy are the most selfish, most willing to use destructive power to harm others to benefit themselves, are the most deceitful and hypocritical amongst us, saying that they help others but are mostly helping themselves so that in terms of the content of their character they are the worst amongst us. This is also consistent with, explains and is justified by Jesus' statement:

"So those who are last now will be first then, and those who are first—politically, socioeconomically— will be last." (Luke 17:17, 18.)

If the very few rich, happy and strong were the better people in terms of righteous content of character relatively to the poor, weak and disenfranchised then Aristocracies, Oligarchies, Dictatorships would be better systems than Democracy.

It is only because the very few rich, happy and strong are worst people morally (i.e. in terms of the content of their character, contrary to the appearances they work hard to present) and the vast majority of the poor, weak and depressed common people are superior morally, that Democracy is the better system.

W. Churchill said that "democracy is the worst form of government except all the others that have been tried." So to be more accurate, Democracy is the least terrible system.

Therefore whether one believes in Democracy and/or believes in Christ Jesus they have to conclude that despite not appearing as attractive, if the judgment is to be made by what is morally right i.e. by the righteous content of one's character, then the right people to be involved with are the poor, the weak and the depressed and not the rich, strong and happy.

If a woman prefers the rich, the "happy" and the strong rather than the poor, the unhappy and the weak then she keeps choosing the wrong men.

Those wrong choices of men, as you know from experience, cause serious problems and damage to others first but then to one's-self.

Many women say that they know that they keep choosing the wrong men but also say that they can't help themselves.

Why do some women want the wrong men?

A part of the answer is that women have to operate in a harsh, deceitful, competitive survivalist system run mostly by rich, strong and famous "happy" men that is designed to entice women to them-selves.

That is why the rest of the book is focused in explaining why it is right, necessary and achievable to establish one small Independent, committed to non-physical violence, Demilitarized Disarmed Democratic Holy State for those blessed who truly value the moral, righteous content of a person's character much more than a person's appearances, as M.L. King dreamed.

The rest of the answer is, unfortunately, profane.

4. Exodus: The migration of the Resurrected

There is order and logic with mathematical precision even in what appears Chaotic, says the Theory of Chaos. What appears chaotic is because it is a fractal, a broken, dimension and as a result it repeats itself through a precise and logical mathematical formula. Most of the shapes of Nature that appear chaotic and irrational, yet beautiful, do have a clear rational mathematical logic and laws that they follow.

By the same logic, if the lessons from the problems and brokenness experienced and recorded in your and our history are not understood and corrected that brokenness is bound to repeat itself in what appears as a different shape but it is in fact the same pattern within the pattern.

Until we learn from history, history repeats itself to make sure we learn because the Universe was (is) Designed to teach.

Because the problems causing The Exodus (that means exit, departure) of the Israelites from tyrannical Egypt over 3200 years ago and the resulting lessons have neither been understood nor corrected they were bound to repeat themselves and are repeating themselves.

If profane things are described by "proper" language, the language deceitfully disguises the profanity of the thing. To describe the profanity of profane things and actions accurately and honestly one must use profane language.

41

The political, economic and social elite, i.e. the strong, the rich and the famous "happy" reject, dismiss and defame as vulgar anyone who uses profane language.

That is not because these major sinners are saints and do not use profane language in private but rather because they are hypocrites, that use "saintly" language in public, while doing very profane things in secret and so they get upset if anyone describes their profane actions accurately by profane words.

Truth is what the deceitful powerful, strong, rich, famous and "happy" avoid the most, and the description of their profane intent and actions with accurate profane language exposes them and so it disturbs them.

Is the primary problem that Christian women keep making consciously or subconsciously the wrong choices about men, contrary to what the one whom they say they believe, Jesus, advised or is the primary problem the systems and institutions that men have designed?

Most of the world, including most of the West, and the US operate in a capitalist system. It is a system of, for and by money.

Capitalism produces significant, money and material advantages over other systems, with the primary opposition coming from a few relatively socialist systems, which in the context of being in a Democracy, can also work well in some cultures, as is the case with some Scandinavian nations.

And there is strong competition from China's dictatorial Kleptocracy, which means the rule by the bigger thieves.

Is the reason that Jane Fonda or Madonna make less money now than Nicki Minaj or Rhianna or the "wrecking ball" woman that they lost their intelligence, goodness and love or is it that because they are older so they can no longer turn people on sexually, despite trying hard with all kinds of plastic surgery?

Do any of these "hot" women do anything else other than try to turn on men sexually acting as if they love us all and want to have sex with all of us?
Is it that prostitutes have sex what makes them bad? Don't spouses have sex also?

So, it is not the sex per se that makes a prostitute "bad" but rather that her intent for the sex is money and is not honest affection or love, as is, theoretically, the case of a spouse.

Prostitutes, through sex, act as if they like and care for someone when in fact they want like and care for someone's money. There are spiritual and there are physical whores. They both pursue money and while physical prostitutes sell their sex and pretense of affection, the spiritual whores sell their integrity and pretense of affection, for money. Both the physical prostitutes and spiritual whores suffer from narcissistic personality disorder.

The *Diagnostic Statistical Manual of Mental Disorders* defines the narcissistic personality disorder as follows:
A pervasive pattern of grandiosity (in fantasy or behavior), need for admiration, and lack of empathy, beginning by early adulthood and present in a variety of contexts, as indicated by five (or more) of the following:

(1) has a grandiose sense of self-importance (e.g., exaggerates achievements and talents, expects to be recognized as superior without commensurate achievements)

(2) is preoccupied with fantasies of unlimited success, power, brilliance, beauty, or ideal love

(3) believes that he or she is "special" and unique and can only be understood by, or should associate with, other special or high-status people (or institutions)

(4) requires excessive admiration

(5) has a sense of entitlement, i.e., unreasonable expectations of especially favorable treatment or automatic compliance with his or her expectations

(6) is interpersonally exploitative, i.e., takes advantage of others to achieve his or her own ends

(7) lacks empathy: is unwilling to recognize or identify with the feelings and needs of others

(8) is often envious of others or believes that others are envious of him or her

(9) shows arrogant, haughty behaviors or attitudes.

To the extent that one values the spiritual more than the physical is the extent to which (narcissistic) spiritual whores are more damaging than the (narcissistic) physical prostitutes. It is the narcissistic spiritual whores who talk without using offensive language but lie and pretend to love you and to do good things for you but want your money and consider you inferior, that give physical prostitutes a bad name.

One may get the exact opposite interpretation of almost any statement depending on whether they interpret it physically, (materially, literally) or they interpret it metaphorically, spiritually.
This same issue in a religious context, of whether to interpret a particular religious text literally or metaphorically is the primary cause of most if not all the religious disputes, conflicts, divisions, and wars in the world.
For example, if a particular religious document says "kill," or "Jihad" as the Koran often does, and one interprets it literally, physically, then one becomes a murderer and the text can be

considered murderous. If that same word "kill" or Jihad is interpreted spiritually which is self-evidently the right way to interpret spiritual texts— unless something is specifically intended to be about the physical— then the conclusion is not murderous but rather is about spiritual regeneration, re-birth, that can only happen by spiritually "killing" the "beast within," the ego.

It is prejudicial and therefore wrong to consider the word whore or a person that is a whore as an automatically bad thing or bad person. There is a great Greek who said: "there are no bad words just bad minds."

Let us consider, let's call her Jenny for this discussion. Jenny is very beautiful but I will not describe her physical beauty because that would give you the wrong impression of her. She was enslaved in the Secret Great Whorehouse since childhood. Jenny still believes in a God that will liberate her one day from the system that she was born in and grew up in and in which she now works. Jenny is kind and gives more than half of what she makes to take care of her mother and gives much of the rest to others. Jenny is very intelligent but she does not believe that there is any way that she can exit the whorehouse on her own.

There are many enslaved physical whores, like Jenny, who live under highly restrictive, oppressive, rules violently enforced, with no way out that they can see, that in my judgment should not be thought of as perpetrators nor be thought of as "bad," even though they are whores, but rather should be thought of as the victims that justly deserve help.

Excluding God, everything and everyone, including whores and bitches, are both good and bad and the how much of each they are depends on each case specifically, its degree, its context, and who judges, by what criteria.
For example, while being a whore might not be necessarily bad, being a big whore is probably bad and being a huge whore is necessarily bad irrelevant of the context.

By that same logic, it is a few bad major beasts that give beasts and bitches a bad name. As it is that a few violent extremist Islamic terrorists give Muslims a bad name.

Consider let's call her Su, who was trying to do her job as a bitch (pimp) in the wh.house of protecting whores from bad customers. In defending Jenny from a bad customer Su got beaten up, shot and is in critical condition.
Now Su says that she was deceived in believing that she was free.
'Believing that I could do anything I want, as they told me, because I could be any kind of a narcissist or any kind of a hysteric or any kind of a borderline personality that I wanted, so that I had plenty of choices among evils, and use my money to buy any kind of thing, so that it felt that I almost had too much freedom, was the deceit of believing that I am free that kept me enslaved into evil, forgive me," she says in her last few moments on earth.
Isn't Su a good bitch and not a bad one?

In translating foreign texts "the rule" is that names don't get translated as to their meaning but rather they are transliterated, that means that a name that makes a similar sound in English is given to them.
Names sometimes don't have meaning in some languages.
Each name and each life have definitely meaning, in Ellenic, i.e. Greek.
When Jesus renamed Simon, with the Ellenic name Peter, he didn't pick a meaningless name.

Because all translations transliterate names, one misses the whole meaning about the character of each of the people of the New Testament and for example, misses what Homer is talking about in the Odyssey if one doesn't know what "Scylla" means. The translation in English of the name Scylla is bitch.

Homer's following passage is from the advice of the goddess to king Odysseus as to how to pass through the deadly straights between Scylla, (Bitch,) and Charybdis:

"There is a large cavern, looking West and turned towards Erebus; you must take your ship this way, but the cave is so high up that not even the stoutest archer could send an arrow into it. Inside it Scylla sits and yelps with a voice that you might take to be that of a young hound, but in truth she is a dreadful monster and no one--not even a god--could face her without being terror-struck. She has twelve misshapen feet, and six necks of the most prodigious length; and at the end of each neck she has a frightful head with three rows of teeth in each, all set very close together, so that they would crunch any one to death in a moment, and she sits deep within her shady cell thrusting out her heads and peering all round the rock, fishing for dolphins or dogfish or any larger monster that she can catch, of the thousands with which Amphitrite teems. No ship ever yet got past her without losing some men, for she shoots out all her heads at once, and carries off a man in each mouth."

Even common people when they call in honest anger some woman a bitch, they do not describe a physical dog, but a territorial, controlling, argumentative, over-judgmental, mean bitch in spirit.

There is the possibility that Homer was nuts given that the probability of such a huge multi-headed physical beast existing in reality is practically zero. The only reasonable interpretation of "huge multi headed beast" is that Homer was not talking about any physical multi-headed beast but was talking of a person that metaphorically, spiritually i.e. in terms of intent, of character content, of thought and spirit looks and behaves so.

Charybdis means whore: *"[A large fig tree in full leaf grows upon it], and under it lies the sucking whirlpool of Charybdis. Three times in the day does she vomit forth her waters, and three times she sucks them down again; see that you be not there when she is sucking, for if you are, Poseidon (Neptune)*

47

himself could not save you; you must hug the Scylla side and drive the ship by as fast as you can, for you had better lose six men than your whole crew."

Homer is explaining, if you understand Ellenic, i.e. Greek, and if you are not a fanatic materialist, how to get through with minimal loses the very narrow gap between an over-controlling Bitch and a greedy money sucking Whore.

Are you stuck between "a rock and a hard place" without knowing who is causing it and how to get through it? I am willing to bet that whenever one is stuck between a "rock and a hard place" there is a control issue on one side arising from a bitch; and there is a money issue on the other side arising from a whore. (I mean these throughout the book of either gender.)

When you get to that "rock and hard" place situation there is a crisis and the only choices left are between pretty bad vs. probably a lot worst. Now that you understand Homer you know how to get through these difficult straights without becoming a perpetrator or a victim.

One gets to pass these narrow deadly straights between the over controlling Bitch and the greedy money loving Whore who sucks, (like a whirlpool) only after one has passed through the sexy singers like the ones mentioned above, the Sirens, which are nymphs, (from which the word nympho is derived,) as was the "hot" Calypso, and other kinds of nymphs, that translated in English, means sluts.
They entertain, by acting "hot" and sexy to turn on sexually others, dumb down their intellect, distract from reality and from the spiritual and deceive in making the violent wrongful money systems sound harmonious and look beautiful.

5. How this world has been ruled

This world and every nation in it is ruled by Money; providing Entertainment in fiction; and using Power-force-punishing to control others.

Do the super- rich through their corporations act as if they like and care for people when in fact they want like and care for money?

The intellectual-spiritual creations i.e. the corporations (corpseorations) have as a primary if not exclusive purpose making money, are accurately described as spiritual whores.

There are many, particularly in business who say that "nothing is free" or "there is no free lunch." Yes, that is true but only because they live in a whorehouse.

Isn't this world ruled by the money of the big corporations; by the control over others through power-force of big government and its law enforcement; and by the intellectual dumbing down, inciting fear, "fiery hot," short-sighted, fictional and deceitful entertainment-news of the big media?

Isn't the entertainment industry, all the movies, all the songs, all the news, about force and violence and usually the more gruesome the better; and about money and luxurious living; and about being "in love" with "burning hot" sex?

Isn't **money** by whores; **power** through bitches/pimps; for the **entertainment** of jerks/sluts what Whorehouses do, how they do it and for whom they do it?
Therefore, isn't this world ruled by a Whorehouse system?

Aren't the systems of men; of the **money** rich, the **powerful** strong politicians, and the happy, **entertaining** for quick gratification media, the systems by which they rule the nations, exactly like Whorehouses, in intent, actions and results?

The rich, the strong, and the (famous) "happy," have been rightly chastised by Christ Jesus because it is they who run the whorehouse systems for their own ego benefit and to the detriment of the vast majority.

The multi-headed serpent like Beast that is described in Ellenic mythology as Hercules's second conquest, Hydra, and by Homer as Scylla, Bitch, are also described by St. John, while he was in Patmos— a beautiful Greek island— in Revelation: "A great red dragon, with seven heads and ten horns" (Rev. 12:3)
"The great dragon was thrown down, that ancient serpent, who is called Devil and Satan" (Rev. 12.9).
Then: "I saw a beast rising out the sea. It had ten horns and seven heads...and the dragon gave it his power and his throne and authority." (Rev. 13: 1).
"Then I saw another beast rising out of the earth.... Let the person who has insight calculate the number of the beast, for it is the number of a man. That number is 666." (Rev. 13:11, 18)

Also, similar to Homer's Whore Charybdis, St. John also writes in Revelation: "Come. I will show you the judgment of the great whore who is seated on many waters, with whom the kings (leaders) of the world have committed fornication, and with the wine of whose fornication the inhabitants of the earth have become drunk." (Revelation 17. 1, 2)

Then, "I saw a woman (the whore) sitting on a scarlet beast that was covered with blasphemous names and had seven heads and ten horns." (Revelation. 17:3).
"The waters that you saw, where the whore is seated, are peoples and multitudes, and nations and languages." (Revelation 17. 15)

"The name written on her forehead was a mystery: BABYLON THE GREAT, THE MOTHER OF WHORES AND OF THE ABOMINATIONS OF THE EARTH.[6] I saw that the woman was drunk with the blood of God's holy people, the blood of those who bore testimony to Jesus." (Revelation 17. 5, 6)

As a result, there have been and there are many modern stories and movies about "vampires," which are a symbol for money thirsty rich piggish greedy whores; and about strong "ware-wolves" as a symbol to a violent, murderous beast in spirit, in the form of a human; and of fiery "dragons" as a symbol to hot sluts.

If anyone is still offended by my use of the word whore, are they offended by the Bible that also uses that word?
Actually, the translation of that word, because the original text of the New Testament was in my native language: Ellenic; that non Ellins call Hellenic or Greek.
But Greek is not a Greek word, it is offensive to Ellins, and there is no "H" in Ellenic, but rather a low octave "e," like in egg. There are many English words of Ellenic origin have that extra but wrongly placed (h) like helix, helicopter, hexagon, etc. to denote a low octave "e."

In common— the Ellenic word for it is "coinoi" — language, St. John is saying that: The leaders of this world rule by punishments, fear through beastly use of force and violence; and by the money/blood sucking spirit of a whore; and by entraining deceiving appearances.

The world is ruled like an oppressive, violent, deceitful, hypocritical, and therefore in Secret, Whorehouse.

Whether one knows enough and understands enough to interpret that this is what Homer taught and what the

Revelation in the Bible reveals, they should be able to understand this:

Any, by definition loveless, system that is ruled by pyramids of power and by hierarchies of money, is bound to operate like a deceitful, because no one in government wants to admit it, violent whorehouse.

If a wh.house is democratic wouldn't that be a good wh.house relative to a dictatorial and thus bad wh.house?
Why would we give a bad rap to the democratic wh.houses and treat them as if as bad as the dictatorial ones?

Our democratic governments are whorehouses but they are much less violent, less hypocritical, less abusive, (and because they are "ours" they are nicer,) than the bad dictatorial whorehouses; to be accurate democratic whorehouses are the least whorehouses and therefore are better.

Some may disagree. Our governments and institutions are not loveless Whorehouses at all, they may argue. But does any government office love you? Of course it doesn't.
The government office will tax you, regulate you, if you are lucky will give you back some of what you gave it, and will imprison you if you deviate any of its thousands of laws, codes and regulations in the name of protecting you or others. Offices can't love; it is not their job to love. If they are getting paid to love, we know what that is called.

Love is Free. Life has been given to you for free. The earth and its air and water and produce have been given to you by God for free.
It is either pimping for control over others and/or it is whoring for money that all government offices do, not loving. (Well, ok. they do a little more; to entertain-inform their groupie bozo flocks they show us their ugly faces in processions, interviews, press conferences, dinners and ceremonies which is their sad idea of being entertaining.)

Christ Jesus, the most forgiving, said: "The ruler of this world has been condemned." (John 16:11)

Why? It is because the ruler of this world is the multi-headed, deceitful, Beast Satan, who has been violently and hypocritically ruling this world, by the spirit of a Whore, as a violent Whorehouse, that hypocritically and deceitfully looks glamorous, "hot" and honorable, through the big happy media of the economic rich elite and through the strong political leaders, Presidents, of the nations.

It is from our democratic less terrible governments that the next stage of improvement must come by giving the freedom for the citizens that so choose, to not be ruled by a whorehouse system at all.

-No matter what statistics one looks at, there is at least 10% of the population in the vast majority of nations including in Western nations and within the US that are in poverty and/or are depressed and/or report extreme stress and/or feel very dissatisfied. The percentage in extreme poverty and/or dissatisfaction is over 50% in some countries.

The current systems no matter how well intended and all the current academic, philanthropic and religious efforts to help these highly dissatisfied, very poor, highly distressed weak, and suffering, at least 10% of the people, have failed and are failing.

That is at least approximately 35 million people, including millions of children in the US that are greatly suffering today, now.

In my view the suffering extends to all the people because each and all seek love and almost invariably fail because they pretend to be virgins looking for true love in a whorehouse and that is asking for too big of a miracle!

Just in case you failed to laugh I put an exclamation mark at the end of sentences that are funny jokes (whether you like them or not) to remind you to laugh!

Even kin love and love within families is having very hard times surviving in these violent hypocritical systems that tear down families.

Aren't all the disputes and conflicts that have torn down most of your relationships and all relationships about issues of **control**, as to who will have their way about to how things get done, and/or about **money** and/or about having found '**fun**' with someone else?

Aren't these the same exact issues by which the violent, hypocritically "happy" and "moral," systems rule?

There are some who preach that love can exist, if you follow them and pay them, within our current violent hypocritical money systems. These preachers are either incompetent or hypocritical and immoral or naively ignorant to be preaching that love can survive within the context of the Institutionalized violent hypocritical whorish systems of our governments.

The meaning of some Ellenic words gets distorted when used in English. One of those words is profane, which in its original means obvious.

In our attempt to not be offensive or as it is called to be "politically correct," we often disguise our words, which can lead to confusion.

For example there was a recent controversy among psychiatrists about the term Borderline Personality Disorder (BPD) not being descriptive enough and carrying some stigma so some suggested "emotionally unstable personality disorder."

Let us consider the characteristic of this personality disorder.

"Symptoms usually include intense fears of abandonment and intense anger and irritability, the reason for which others have difficulty understanding.[1][2] People with BPD often engage in idealization and devaluation of others, alternating between high positive regard and great disappointment.[3] Self-harm, suicidal behavior and substance intoxication are common.[4]"

The World Health Organization's ICD-10 defines it as follows:

F60.30 Impulsive type
At least three of the following must be present, one of which must be (2):

1. marked tendency to act unexpectedly and without consideration of the consequences;
2. marked tendency to engage in quarrelsome behavior and to have conflicts with others, especially when impulsive acts are thwarted or criticized;
3. liability to outbursts of anger or violence, with inability to control the resulting behavioral explosions;
4. difficulty in maintaining any course of action that offers no immediate reward;
5. unstable and capricious (impulsive, whimsical) mood.

F60.31 Borderline type

At least three of the symptoms mentioned in *F60.30 Impulsive type* must be present [see above], with at least two of the following in addition:

1. disturbances in and uncertainty about self-image, aims, and internal preferences;
2. liability to become involved in intense and unstable relationships, often leading to emotional crisis;
3. excessive efforts to avoid abandonment;
4. recurrent threats or acts of self-harm;
5. chronic feelings of emptiness.
6. demonstrates impulsive behavior, e.g., speeding, substance abuse[78]

This is a well-reasoned, precise, scientific, non-offensive, complex, long, politically correct definition to avoid the stigma of what common people call a slut.

The following are the criteria according to The Diagnostic Statistical Manual of Mental Disorders for Histrionic Personality Disorder
A pervasive pattern of excessive emotionality and attention seeking, beginning by early adulthood and present in a variety of contexts, as indicated by five (or more) of the following:

(1) is uncomfortable in situations in which he or she is not the center of attention

(2) interaction with others is often characterized by inappropriate sexually seductive or provocative behavior

(3) displays rapidly shifting and shallow expression of emotions

(4) consistently uses physical appearance to draw attention to self

(5) has a style of speech that is excessively impressionistic and lacking in detail

(6) shows self-dramatization, theatricality, and exaggerated expression of emotion

(7) is suggestible, i.e., easily influenced by others or circumstances

(8) considers relationships to be more intimate than they actually are

So, if one wants to avoid as much as possible saying the obvious, the profane, the self-evident, explicitly to avoid offending others one may use the Psychiatric terms to accurately describe spiritual and physical whores as narcissists, bitches-pimps as hysterical with histrionic personality disorder, and sluts (of both genders) as having a borderline (schizophrenic) personality disorder.

All these are character disorders; they show really bad content of character, a dead conscience, that over the long run make the color of skin and other appearance issues relatively irrelevant in comparison.

That is why I have used "happy" in quotes because even though the famous entertainers who usually suffer from borderline personality disorder appear happy, they are pretending; in truth they are usually very sad and are trying to escape it; as we keep finding out by their tragic scandals, suicides and their by intoxication caused deaths.

The spiritual suffering caused by the emptiness of superficial relationships, the loneliness, depression, anxiety, fears, confusion, coveting, shame, anger, hate and personality, (character) disorders, characterized by a horribly ugly character and/or a dead conscience that is absent of righteousness, is across all societies and classes.

Is it right to glamorize or idolize the mentally stuck in childhood "athletes" who are adults that still keep chasing balls?

Is it right or correct to glamorize your scholars and politicians as people who "know" when in fact if they really knew they could predict, while their predictions are always wrong? They are false prophets, and they can't even predict who their spouse will have lunch with next week.
Is it correct to glamorize and idolize the false idols of happy looking but sad ignorant actor/ actress who pretend and lie for a living?
Is it right to glamorize gruesome violence and deaths and destructive cunningness, in the name of art?

Admittedly some actresses/actors lie beautifully. But since when someone being a convincing liar, is the right reason to glorify or idolize them, as the bad immoral short-sighted sick sad attention seekers of the big media do?
Is this a culture that prefers lies and liars to honesty and the honest?

The biggest obstacle for the liberation of the slaves of the violent Secret Great Whorehouse, that is this world, is that too many of the slaves like being slaves to deceitful, hypocritical and morally bankrupt masters.

Whether the primary cause of suffering of humanity is that women knowingly choose the wrong men— i.e. the strong beastly wolf-like pimps that are the professional "athletes," the police, the politicians and lawyers; the rich corporate vampires- whores; and the dragon like "fiery-hot" pretty looking short-sighted sexy nymphs that are the entertainers and the Media —or the primary cause is that the systems of governments by men, are deceitful hypocritical violent Whorehouses, both have to change to at least some extent for a lasting change.

I was depressed and isolated for over two decades, even though there was nothing bad physically, economically or materially going on in my personal life that I could reasonably complain about. Externally, it is the News that depressed me most. We all know what the news will say; we just don't know the specific names, places and times.
But once you fill in those and look at all the local, national and global daily and monthly news of the world, this is their summary:

THE NEWS

Somebody was shot … someone got injured… someone got raped… somebody was strangled,…(we'll tell you who after you look at this ad) and somebody drowned… (look at these t.ts…

and buy that). This many were bombed and died in the war in the Middle East.

Somebody was knifed, several were burned... someone beheaded, someone got lost and someone got beaten up. Politicians called press conferences and made speeches about those events and said that they want to pass legislation, and that they are serious about it this time, (look at this ass and call this number for a great deal for some product that has nothing to do with this ass.)
The President said in an interview with our own reporter that things are getting better.

Hundreds ended up dead in a crash. The Muslim terrorists issued a report that they are satisfied and will win and then they blew themselves up.

A new amazing movie with the super-sexy .... (here's her picture with a good side view of her t.ts and thighs all the way up)... shows a whole city in shambles.

In sports news, our city's Baloneys beat the lousy city's Salamis, decisively. Here are highlights of them running and cheering; and the losers walking out as if they just got screwed badly; the expert said that it's because they weren't keeping their eyes on the balls; and here's our panel's babel about it.

Our sports athletes have proven again that the purpose in life is to get the most balls in the holes. The athletes that get the most balls in the holes are rewarded with the best vaginas, because that is what life is all about. (Here a video of a sports athlete being interviewed by a pretty vagina describing how he got the ball in hole despite duress and opposition.)

Breaking news; somebody else was shot and this many were bamboozled....(look at these very happy people dancing because they bought this...call now... and don't miss your last chance)....somebody overdosed, someone was blown up and this many were found dead.
Now... something a little more bizarre... that we will obsess about for as long as we can...

The weather: the North East is cold with a new bad snow storm; the South West is dry and burning up on fires; the South East was hit by a tsunami; a hurricane with severe flooding and tornados hit the South, and a volcano and an earthquake shook up the North West. Here are the scenes…Enjoy or cry by the scenes of people suffering through them.

The medical news is that there is a new epidemic. (Do you feel safe and pretty? If not buy this… just for Jesus' sake this is the only time of the year that you will not feel very badly screwed; hurry.)

Special Report: Large numbers of people got robbed, punched, kicked, slapped, bullied, sexually harassed and then arrested and jailed because they were poor and did not have expensive lawyers. A whole bunch of people got lost today and can't be found. Only a few lost a limb today.
The President underwent a free colonoscopy; his prostate was also double checked, and is fine.

We will repeat the daily news that we have been reporting for over two centuries and show you another debate of a Republican saying as they have for over two centuries that they want less taxes; and a Democrat saying that we must spend more. So enjoy another 2 bickering matches today, arguing the same arguments since G. Washington retired.
Today we have a new subject and new faces of Senators spitting on each other the same arguments since the 30 year war of the Democratic Athenians and the Republican Spartans.
Now, watch our politicians solve the problem by cutting taxes and spending more and so making your children carry a huge debt.
Let us see if we can find some ass that doesn't piss us off too much and elect them; we will interview all the asses for you and then you judge as to whom you hope will f. you less.
Former someone offered an Analysis: The majority, over 4 Billion humans remain clueless and starved (to reduce how guilty you feel and are, send them a package of crap…for only 99c.)
And now for something funny: our educational system is still for laughs.

The President of country X met with the President of country Y, and agreed to boycott and then bomb country Z.

The war in the Middle East is thriving with people hating each other more than ever; here are pictures of recent carnages.

Lots of scums are still trying to f. and rule Greece that is still barren and beautiful but broke and in debt, as for the last 8000 or so years, when the Greeks of Atlantis learned the hard way to not mess with Elli, when the volcano of now beautiful Santorini put them all on ships, as did the eruption of 3600 years ago, and they are still on ships fleeing Greece but there are some positive economic statistics that in 200 years things might get better. Children in Ethiopia are still dying from poverty all over the place. An Ethiopian and a Kenyan won the marathon today. They attributed their success to the chicken they had to keep chasing for dinner as children.

There is a hostage crisis right there. A few species got wiped out today. We were hacked again today. We are not allowed to report on suicides but their number exceeds the 60 or so murders that happened in the US today. There was a big gala party in the capital where all the dignitaries enjoyed themselves. Mrs. Fme wore an outrageously expensive outfit, designed by the well-known gay man Ou la La, which shows her thighs and tits in a very classy way. (Here's a picture.)

Somewhere, people protested and we ignored them but those we noticed were gassed and chased away. There is an important debate in Congress going nowhere fast…more on that after the next elections.

A very important person said something that sounded profound but was junk!
The good news: The statistics are in and they are excellent; only 70 women were found to have been raped today in the US. Bill Cosby and the National association of sugar daddies hailed the reduction of illegal rapes with a nice party with caviar and everything, with all the dignitaries, in which no profane words were used when doing very profane, immoral and illegal in spirit, acts.

Following the example of our Greek friends, Americans are very reluctant to vote for anyone who doesn't have a "big name" that became famous by screwing the Country really badly, so our candidates for the next President is another Clinton and another

Bush, because neither family feels that they did enough yet to completely screw and bankrupt the nation, as the two big names of Greek families successfully did.

Countless people got screwed last night for and by money. (Our talk show provides expert analysis regarding which ignorant bozo said what; right after you buy this.)
Thank you for letting us into your home. Have a lovely day!
(Don't miss our special tomorrow at.... about nothing, with new and improved happy lips, rouge, make up, and brand new plastic tits reporting on it!)

Why do we still keep looking at the News for something new?

Now watch our regular programming of the acclaimed Award winning series... and Award winning actresses containing obscene degrees of violence, deaths, luxury, profanity and graphic sex but because we are "good and moral people" we will not show the vaginas. (For viewing of the vaginas also, magnified and in HD, give us the number of your credit card!)

The routine during my many years of dark days was to wake up in the afternoon, ask God to kill me now and end my misery, look at some News, confirm to Him that this is a beautiful day or night to die, write on some subject cussing my brains out making rappers look timid and then to go back to sleep.
I prayed that God finally listens and stops waking me up to convey His message that I have been failing to do so far maybe because I have five grammatically challenged languages mixed up in my stupid brain.

Now that you know all the news from all the countries and at all times until the change that is recommended in this book happens, do yourself a favor to avoid depression, as I did, and stop watching the news or at least watch them on "mute."

6. Understanding the reasons and the value of both opposing sides

Because there are two sides to every story, one can of course construct arguments that our governments and this world are not wh.houses and that none of the above is true, as far as they are concerned.

From a physical point of view, in terms of appearances, there is no statistically significant higher physical whoring going on in governments than anywhere else; whoring must be physical according to its legal definition and includes quid pro quo, is mostly illegal, so that the assertion that governments are wh.houses is not just factually wrong physically speaking, but it is a false accusation.

Since the invention of credit advanced whores have moved to strategic quo pro quid whoring, meaning first the deed and then the payment. Also, as you might be aware from the porn industry, high tech physical whoring doesn't even require physical whores to be in the physical vicinity of, or to physically touch their online customers. But aren't "porn stars' still whores?

Democratic-capitalist systems despite their very many flaws are the best we know of, for generating and diffusing wealth and power, decreasing poverty, reducing wrongs and defending against the greater evil of the dictatorial governments that allow much less freedom.

How one perceives and interprets life depends on their own state of mind, specifically whether they are materialists, realists, pragmatists and thus Aristotelian philosophically; or are idealists, spiritual and thus in Plato's world of ideas. I do not deny the value of either approach.
I intend to show that both are valid, true and valuable.

Even though they seem opposite, both answers to what in philosophy is called the ontological question i.e. whether ideas come first or matter comes first; are both valid and true. Here is why:
From the prospective of Ideas, ideas come first; and from the prospective of matter, matter comes first.

That is because each thing, every issue and every person and organization has two sides to it that are: what it is; and how it is perceived, i.e. its image, which is the appearance that is a perception and an idea in someone's imagination.
There is "is" and there is "is' image" both being there at the same time.

In that sense everything and everyone (except God) is what a physicist would call "dual."

If one believes themselves to be a thing, they are a thing and matter; so Mother, and Mother Nature and appearances is what matters to them most. If one believes themselves to be a spirit, an eternal soul, living in the world of Ideas and ideals, the shadows of which we see as the physical, as Plato argued, then ideas the moral righteous content of one's character and the Spirit of the Father God Elli is what matters and should matter most to them.

The existence of both ideas and matter is true; and both opposing sides are needed for pro-creation and for development and progress.

If one examines life and the Cosmos as if it is only solid matter then the answers are mechanical and deterministic, i.e. with no real choices, so if one accepts that they really have no choices then a materialist fatalistic approach is valid.

To assess any threat or any damage or any problem one needs to examine the material facts whether one claims to be a materialist or an idealist.

If one wants to create solutions, believing that they do have free will and choices to any of the problems in life, they need ideas first.

Whether one is a materialist, realist or is an idealist, spiritual, reading this or any other book is evidence that one is looking for the ideas that can cause positive change to whatever problem one is having or is thinking about.

Because thought itself is a not necessarily a directly visible physical phenomenon, to answer any question, to solve any problem and to create, one needs to operate as an idealists irrelevant of what they claim to believe in.

In simpler language: To understand what is going on, you need to be a realist; to change and improve what's going on, you need to be an idealist.

Both sides of every argument, dispute and disagreement have some validity and value and truth. Why? In the final analysis there are two sides because both Light and Darkness are true, are real, exist; and both truth and lies are true; they happen.

The question, after what is true, is what is right?

A mathematical and geometric way to understand how both an Aristotelian, realist, materialist; and an idealist, spiritual Platonic philosophy can be both valid and true yet only one of those is right, one may define the reduction of physical wrongs as a reduction of negatives and the increase in righteousness as an increase in positives, in a straight line.

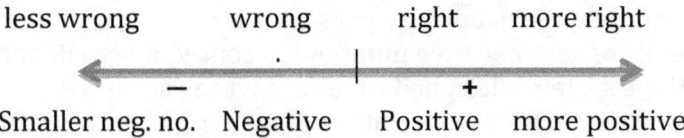

less wrong   wrong   right   more right

Smaller neg. no. Negative   Positive more positive

Both reducing a negative, a wrong, and increasing a positive, such as righteousness are valid and worthwhile purposes yet they require introduction of opposing methods, opposing tactics, are in direct directional opposition and only one of them is right.

Reducing or transforming a negative number requires the introduction of another negative number. Increasing a positive number requires the introduction of another positive number.
Similarly, reducing crime often requires penalties for the crime enforced through violence by the judicial system. On the other hand increasing the well-being of people requires rejection of violence and the increase in righteous acts.

The arrow of time and of history leads to increased freedom. It is not increased concentration of power and control but reduced concentration of power-wealth through increased freedom that produces the reduction of wrongs.
Because of increasing democratization throughout the world, there is much improvement in the freedom, the honesty and the material condition of humanity by most statistics with increasing average age, reduction in abject poverty and of crimes, wars, and so forth.

The other side of this is that while the degree of wrongs has been reduced the morality (righteousness) of humanity has deteriorated.

The legal definition of whoring is physical, because if it included spiritual whoring, like for example, depending on who is paying a lawyer, the lawyer arguing for the guilt or innocence under the law of someone, then almost all the lawyers would be in jail.
(See also Luke 11.46, about lawyers.)
An example of very negative unintended consequences in our competitive systems designed to reduce wrong as our systems are, is that those who attempt to do right, and therefore give without taking, are taken advantage of, get abused and at best they cause co-dependency. "No good deed goes unpunished" (in this world) they say.

A key component of succeeding in these systems that is frequently overlooked is that while it is true that one's strategy should be based on one's strengths; the organizational structure is driven by one's passions and is based on one's weaknesses.

The word passion originates from the word "pathos" that means (emotional) damage.

You may have heard CEOs such as Bill Gates and Andy Grove (co-founder of Intel,) say that "only the paranoid survive in this business." So, being paranoid, which is a negative personal attribute, is very useful in business.

When I was running businesses it was the weaknesses that I found most useful in people. Someone was timid and scared, I saw it as excellent; someone was screwed up, I saw it as great; they were angry, I saw it as useful; they were confused, I saw it as beautiful.

All weaknesses and passions so long as they are correctly identified and correctly re-directed separately, are essential to generate drive and useful energy.

For example, in my opinion, it was the horror from Saddam Hussein's assassination attempt against H.W. Bush, after Bush senior left office, that drove the son G.W. Bush to the Presidency and to the second Iraq war, to rid the world of "evil Saddam." (No other rational for invading Iraq makes any sense.)

In my view, it was the "weakness" i.e. the great pain and anger of losing his mother while "she was dying with cancer and having to keep arguing on the phone with her insurance company" that passionately drove Obama to the Presidency and to Health Insurance reform.

That is why getting rid of Health Insurance Reform while Obama is President will not happen; improved is the best that can be done. To the extent that Republicans keep attempting to get rid of Health Insurance Reform they only show malice to the other side.

Another example where correct identification of weaknesses is critical and useful is that if for example a CEO is strong in Finance and weak on Marketing they should not be hiring based on their strength someone good in Finance but on their weakness, someone to complement their weakness in Marketing. So, while strategy must be based on strengths, the organizational structure must be based by hiring around weaknesses, and it is weaknesses and passions that drive people and organizations.

This is consistent with the judgment above that it is the most screwed up in terms of character who have incurred serious moral damage and thus are most passionate who then combine that with the hypocrisy of talking nicely and appearing nice that become your leaders in this world.

So both strengths and weaknesses can be significant contributors to one's success and to a system's success, if and only if they are understood and directed **separately** into separate contexts correctly.

Once one understands the distinct value and validity to both opposing sides similar to the opposition of reduction of negatives vs. the increase in positives, then one recognizes that most of debates, disagreements and conflicts can both be well intentioned and can complement each other but fail and result in people damaging each other needlessly because of lack of understanding as to how to **separate** the contexts in which to apply the opposing tactics.

For that reason a **separate** system can, needs to and should be built to complement the weaknesses of the current competitive whorish for money, violent for power hypocritical systems ruling this world.

Both philosophical perspectives of idealists and of realists are valid and valuable but because they result in the necessity of opposing methods and actions, as has been shown by the straight line above, (unless one controls the phasing extremely well,) those who believe in the material, physical and thus in the reduction of wrongs, which is what the governments and our systems do, and those that believe in the spiritual and thus in the morally right, **must be separated** into Independent societies by Independent governments or else their efforts even if well-intentioned counter each other uselessly, damage each other unnecessarily and both sides come out losers.

Not separating these two groups, is as wrong as letting endless bickering and bloody deadly fights go on and on, without separating away from each other, those who swear that they will never agree with the other.

I do not at all deny the value of our current violent competitive whorish systems nor do I recommend replacing them.

There are many, probably the majority, for whom these systems work well and they have, in the past, worked well for me also.

Maslow's hierarchy of needs shows that once the material, physical needs are met, there are higher emotional, social, intellectual and moral needs that humans pursue, such as love and such as being true to and actualizing one's higher True Self.

Consistent with Socrates' theory of opposites: that everything in the Universe comes from or through its opposite— and is viewed in relation to and in the context of its opposite— it is the material success of our democratic-capitalistic competitive systems that allows us to now consider and to have an additional not competitive but a cooperation based system not to replace but to compliment the current system so as to meet the higher human needs such as the need for love.

Because of that principle about how life operates bringing also the opposite of what is pursued, unless the imbalances keep getting correctly re-balanced fast to stay on the straight path, —some call it the Middle Path or Path of moderation— we keep finding tragedy and irony in life.
Central unsustainable imbalances are the wealth and power imbalances, which have gotten to unacceptable levels and only the solution that follows will restore them fast enough in a way that is beneficial to all.

Because there are some, even if a few, as few as 10% who actually believe and view this world and these governments as ruled by the most immoral, selfish, greedy, hypocritical double talking and double dealing whoring and beastly spirits, one, in a democracy that is intended to give every citizen, including those that are in the minority real choice, must offer these few 10% or so the choice of not being bound and enslaved by this system.

There is a significant and valuable minority, in my view the most valuable humans, who view this world spiritually and as a result they see this world as a violent hypocritical Whorehouse, and they include Homer, Socrates, Plato, St. John, Jesus Christ, the apostles, the saints, and the prophets. By spiritually I mean looking at the intent of the person(s) and using the physical as metaphors to describe the intent and the (opposite in appearance) result.

Recently a few bad apples say the same thing based on a book that was written long ago by an Egyptian about America; most Islamist terrorist believe and say that America is a Whorehouse.

To the extent that winning the "war on terror" includes defeating not just the abhorrent, condemned methods of terrorists but also their ideological prospective then proving by actions that our government is not a just another violent deceitful enslaving into self-interested wrong actions and thus into sin Whorehouse is central.

So, that letting the poor, the unhappy, the weak, the disenfranchised, alienated and victimized, such as the kind prostitute Jenny and the brave bitch Su, have the option of a separate kind of government and system that is Independent and is different from the one that is causing them the suffering, is centrally critical not just for the spiritual but also for the physical well-being of both philosophically opposing sides of humanity.

Any competent executive knows that when they want to invent and make something quite different and separate, they need a separate organizational structure for it.

Islamist terrorism against the West will never be defeated without such a change within and by our own governments.

Even if one likes their violent, for and by money hypocritical government because it is theirs, on the basis of an animalistic group and groupie mentality that is unfortunately widespread these days that if it is ours no matter what it is, it is good (and thus we call it by a nicer name, such as a democratic-capitalistic system) should one commit one's children and future generations to violent hypocritical Whorehouses or not?

Why would anyone deny children the opportunity to live in a non-violent culture where they don't have to be getting bullied or to bully?

The central issue then becomes, let some exit the whorehouses, fine, but to go where?

If one thinks of government as a Beast, like Homer described with "three sets of teeth" such as the national, regional and local government regulations and taxes and arbitrary enforcement of thousands of laws, then getting out of the influence of the Beast requires a fully Independent State, Republic.

As an Ellin, I feel to a large extent responsible for our competitive systems, under fair rules ethically implemented, i.e. for Democracy and even for capitalism (for example, the ancient "agora" means and was a free market) so long as one recognizes that there is nothing to brag about "being in the market" so that you may blame the Babylonians and the Brits for glorifying capitalism.

However, to believe in any single system in which people get paid to do things and is therefore loveless, so much as to force everyone in a nation into the same single system is offensively stupid.

Understanding the value of both of the externally conflicted sides is important because they will help you understand both of the conflicted sides internally, within you.

Truth exposes the hidden which sometimes happens to be ugly. So, I know that the truth feels harsh. Just like light feels harsh to look at directly.

As a result, many have trained themselves to avoid looking at the truth or telling the truth.
What is your personal experience with the strong of big government, the rich of big business and the happy of big media? Do they really care for you as they keep pretending? Are they really doing what is right for you?
Am I lying? Am I insane or irrational?

The truth is the least harsh way to effectively deal with the problems of reality.

Jesus said: "Everyone who commits sin is a slave to sin." (John 8. 34.)

And: "If you continue in my Word you will know the truth and the truth will set you free." (John 8.32.)

7. The Only Right Solution for Both Opposing Sides

The first central issue that needs to be done differently for the weak in the New Independent State is to eliminate violence by the "strong."

There is a hypocritical contradiction in preaching peace and non-violence on the one hand; while living in systems where violence under certain circumstances particularly by the government, by the military and by the police is not only considered acceptable but honorable.

Those who honestly claim to believe in and want to live in non-violence need a government that doesn't use violence under any circumstances either. Government officials talking about non-violence while they are one the biggest perpetrators of violence, is hypocrisy.

Therefore, the citizens of the New Independent State must be committed to non-violence, to no physical violence against any other human, including by its government. As a result the government must be demilitarized and its citizens must be disarmed.

For this to work the New Independent Demilitarized Disarmed State must have and exercise the right to exile any of its citizens who violate its laws and who are not forgiven. Exiling citizens from the New Independent Demilitarized Disarmed State must by Treaty and must be implemented by one of the currently militarized nation States so that the disarmed government of the New Independent State doesn't have to use force.

But that is not enough. The commitment to non-violence would eliminate or reduce the system being a violent wh.house but it is still not enough to eliminate it being a wh.house.

For example, after WW2 Germany was demilitarized and that was critical for its restoration and for its great economic growth since. So, now Germany is, again, at least a not too violent anymore, corporatist huge bad Whorehouse dominating and abusing economically the Eurozone by "tyranny by money" which T. Jefferson called "the most insidious tyranny of all."

For example, Greece was doing quite well economically prior to entering the Eurozone, yet soon after it entered the Eurozone Beast Satan's evil huge whore A. Merkel intentionally put Greece into a 6 year Depression resulting in over 25% loss in GDP and increased debt that the Greeks were told they had to take at usury rates and in dramatic increases in suicides, in poverty and of children in poverty. If that isn't mass murderous extortion, including by a 3 week closure of all Banks in Greece, of a whole nation, by an "ally" what is? Either Greece or Germany must and eventually will probably exit the Eurozone.

When the debt crisis in Greece started in 2010 as a result of the US and German Bank crisis in 2008, the Greek debt to GDP was unsustainable at 120%. After the unethical adding of loans to unsustainable loans, by "bailout loans" from the "allies/creditors" it is as of 9/22/2015 the surely unsustainable over 180% of GDP.
So during the next bailout loan Greece will need, probably in 2018-19 there will be a Greek Exodus or a German exit.

(I hope that during the 11/2015 negotiations at least the average interest rates will be reduced by .8% to 1%)

Merkel's arguments of a write-down being inconsistent with the Eurozone regulations is false because there is no regulation that says "no write-downs" yet there is a regulation that says "no bailout loans" which is being violated blatantly and repeatedly.

The "moral hazard" argument is also false because neither Italy nor Spain (which Merkel wants to threaten from asking a write-down also,) would be willing to get impoverished, with more debt and enslaved as Greece had to for 7 years, to get a "haircut."

It is a natural, honest and a correct conclusion that the world is ruled as a violent Whorehouse, not just spiritually speaking but also from a realistic, materialistic, pragmatist point of view.

Darwin was correct in identifying that from a materialistic, physical point of view, sex is the central mechanism of procreation and survival of a species, and the dominant role of women in choosing mating partners is discussed in his publication *The Descent of Man, and Selection in Relation to Sex.*

Freud was also correct that from a physical, materialistic point of view sex is a, if not the, central motivating force of adults.

So that the pursuit of money, strength, fame by materialistic men is not for spiritual reasons but for sex from women; and the pursuit by women of beauty through competitions in beauty, by singing, fashion, dancing, make-up, through plastic surgeries and everything else are not for "liking themselves" purposes but they are to promote "higher quality sex" for men who offer more money, strength, fun and good reputation, fame.

Therefore the spiritual whorehouses that rule the world have manifestations of physical sexual whoring that is obviously done in secret because of correctly felt shame in whorehouses, in hotels, through "fun trips," in bars or hypocritically at work with a "boss" or a "mentor," in bedrooms through pornos, online, escort services, "3rd dinner dates," gifts or through having relationships or marrying for money and divorcing.

I personally evidenced how much "cuter" I seemed to appear to women when I showed off and distributed money, than when I didn't seem rich.

By this broader definition of whores, where payment for the sex is understood to happen and happens at some point but is not explicitly negotiated (just as political contributions even though don't have a specific payoff it is understood that there will be some repayment at some point) there is huge amount of physical (and spiritual) whoring going on in every nation and every city.

To the extent that Freud and Darwin are correct that from a materialistic, physical "survival of the fittest" point of view we evolved from animals-beasts-germs and are driven by sex, it would be unreasonable to conclude anything else other than those through animal evolved humans are bitches, sluts, whores, pimps and jerks, as all animals are, cute as they maybe and that they are operating or are being operated by sex in violent whorehouses, spiritual and physical. And, by the logic of "big fish eats small fish" the strongest, most cunning and most deceitful would be the leaders of these survivalist competitive hypocritical violent but glamorous looking Whorehouses.

Wh.houses do not co-exist peacefully with each other; they compete against each other for survival and at times their disputes end in violent conflicts and wars.

So this world is operated by deceitful violent hypocritical wh.houses in competition and sometimes at war with each other.

In that sense, the wars of humanity are wars of, by and for whores.

As Homer explained in the Iliad—which is historical rather than spiritual as is the Odyssey — even though one may find some who have merit on both sides, both sides are wrong when in war.

Human wars are wars by pimps, beasts and narcissistic male and female spiritual and physical whores for "beautiful" but short-sighted and faithless (untrustworthy) borderline schizophrenics, like Ellen of Troy.

The government of the New Independent Demilitarized Disarmed State would be much better if it is Democratic and even better if it has "the least government" because A. Lincoln was correct in saying that "the least government is the best government." And the worst government is the "totalitarian" biggest government.

These rich, laughing and "strong" believe, (must believe to get there,) that they are better than others and by controlling, impoverishing and saddening others they feel that they are "better." But they are morally the worst and in time they will find out and pay for it.

The morally better, in the context of wrong, are the least strong, i.e. the weak, the least rich, i.e. the poor and the least "happy" i.e. the mourning, which is why they have been blessed.

In the context that Darkness pre-existed Light,—which is valid both scripturally and scientifically— which is the necessary imperfection and evil within Creation, to be correct one has to add to the above statements of "least is best," "least but as much as honestly, objectively and not cowardly necessary to survive, is best."

Finding and maintaining that narrow straight balance physically and spiritually i.e. morally, intellectually and emotionally, is not easy but it is the best path in the context of the existence, the pre-existence, of at least some evil in this world.

There is a lot of evil in this world these days, so that understanding correctly what those straight balances are that allow for highest speed, highest growth with the least risks, is crucial.

With beautiful poetry, rhythm, balanced symmetry, music and melody that are all lost necessarily in all translations, Homer explained how to get out of all "rock and hard place" situations, exactly.

To start with, there is a gap between the whore and the beast; between the control issue and the money issue that cause the "rock and the hard place" and it is enough of a gap to separate the issues of money from those of control because the motives and methods of both these evils, (the whore and the beast,) operate differently.

Let us take for example, the stalemate between a "rock and hard place" that the politicians of the Democrats on the "left" vs. the Republicans of the "right" find them-selves in, these days, and through which they put this nation.

It is clear that the political "left" is into more control through bigger government and so that in this context they are the Beast.
Politicians on the "right" are clearly into growing the economy and are about money so the political "right" is the Whore in this context.

Homer said to not take the arithmetic middle of the road, but to get closer to one side, to the side that has the lesser risk, the side of the bitch.

To be precise one should take the path of the "geometric mean," that is also called the "golden mean" because the two sides aren't on a straight one Dimensional line really but are two separate orthogonal Dimensions intersecting. The geometric mean expresses the proportionality of nature's beauty that in percentages is around 62% vs. 38%.

Homer also explained that you also have to wait for timing that is not wrong timing. Once you determine the path, by 62% closer to the safer side of the Beast, you have to move fast and go straight while remaining in straight balance.

The differences in the various budgets proposed by Democrats vs. Republicans are about + or – 5%. Politicians deceive in calling their + or - 5% of the money differences "ideological." What is "ideological" about a 5% of the money difference?

So when politicians on one side argue about a specific number for a part of their fiscal or other policy and the other side has put their "best and final" offer and there is a gap, if there was any truly Wise person, s/he would pick that geometric mean between those two numbers, favoring a bit the beasts on the left and close the deal fast.

That is the straight, narrow, less traveled, honest, beautiful, harmonious, best balance, highest growth path.

The most serious problem and cause of failures of any for profit business, non-profit organization and government is the lack of a cohesive and straightly aligned strategy. As a result much of what is done is self-contradicted and therefore both ineffective and inefficient.

The straight alignment that is necessary for success is the alignment of the mission/purpose with the objectives with the strengths based strategy to achieve those objectives with the weaknesses based organizational structure, including the alignment of incentives and disincentives of people with the goals, with the tactics, with the measurements for performance and with the process for rebalancing after each phase and reprioritizing for straight realignment.

That is the Straight Path of Jonas, Aruna, Odysseus, Homer, Elijah, John the Baptist and John of Patmos who "make His Paths Straight." (Mark 1:3)

A central obvious misalignment between what people say they believe in and pursue vs. what they do is that almost all who have authority are significantly invested in evil. If God showed up and miraculously ended all wars for good, would He be hailed by the defense industry or despite saying that they are happy about it wouldn't they do all they can to not go out of business? And how happy would the doctors and pharmaceutical and insurance companies be when finding out that they have no money to pay for their big mortgages because disease was eradicated?

How badly would lawyers persecute God if they found out that they are out of business also, and can't charge anymore $500/hr. for saying bad baloney because there are no legal disputes either any longer? As shown, most big businesses are invested in evil and have the financial incentive of increasing evil even though they say that they are fighting it.

What these people say they believe in, the goodness of the image they project and what they are financially invested in are in direct opposition; are clearly misaligned.

Homer also wrote that one, very unfortunately, has to lie to flatter and keep flattering the Beast, while going through the "territory" that she "controls" to keep her appeased.

If poor Homer, who is from my beautiful ancestral island of Chios— in which the descendants of Homer have become servants of an exciting summer camp for German sluts— knew that he would be ignored by his generation of Greeks and it would take almost 3000 years for the people who "honor" him to understand and practice what he was teaching so that they get out with minimal harm between the rock and hard place situations that they still get into, he would probably say: 'screw it, it is not worth explaining it.

Let me get drunk and marry and let some other bastard in some future generation do the educating of those incapable of even recognizing that they have been barbaric and let someone else do the enlightening of the spiritually blind and the providing of loving help to the barbarians and the Babylonians.'

That is what I think poor Homer would say, as Prometheus would, if they knew what a thankless Sisyphean job their descendants were getting into by trying to civilize whorish Babylonians and beastly Barbarians and the huge price that future poor Greeks would have to keep paying for it.

It would be much preferable if money and the economy are a minimal factor of what the government of the New Independent Demilitarized Disarmed Democratic State is about, if it is not going to become wh.house.

To achieve that, the differentials of wealth and income must be agreed upon upfront to be much lower than those that the current systems have.
There are simple "system" ways to achieve those objectives.

Because these new Independent demilitarized governments, in my view, must be democratic, what economic system they have is up to them.

However, here are some suggestions:

-   The money-currency.

To be truly independent these new demilitarized, disarmed States will need to have their own currency. The currency should probably have a serious name but for these purposes let's call it, "the baloney," as any virtual currency is. For foreign exchange purposes one baloney may be linked on a one to one basis, to any other kind of other baloney currency such as one Yuan, Euro or dollar.

However the baloney should be a more serious currency than these bad baloney currencies in which most of what you think you own, is owed by your government.

So to avoid the baloney currency of these new Independent States, becoming bad baloney, their government should not have any debt.

(Ellin) Mystics wrote on the dollar: "In God we Trust." The problem is that it is hypocritical now because most "in money they trust" and God has already said that He, in money doesn't trust: "You cannot serve God and Money." (Matthew 6:24)

Governments serve money and anyone in them claiming to be serving God also, lies. If any did serve God they would at least give the option to those suffering under this system to leave it, have their own Independent Demilitarized Disarmed government while remaining part of the nation.

- The ownership:

Any public corporation to do business in the Independent Demilitarized, Disarmed Democratic States must give at least 20%, within 5 years, of its local ownership to non-executive employees in some stock option plan.

Aligning the incentives of everyone is critical for success.

- Wealth distribution.

Most governments, including the US Federal Government, spend over 20% of GDP. Given the reduction of government expenses by avoiding defense expenses, interests on debt and all other baloney program expenses which are bad "crapitalist,"— as an interesting reporter's book exposes —excuses for redistribution that is not really going to the poor as it is presented but to the politicians and the rich who fund the politicians, government expenses can be cut by over 50%, allowing for significant increased growth potential with much better wealth distribution. For example:
The tax system could be highly progressive, simple to remember, collected automatically online and the

redistribution could also be automatic and online to avoid the current huge inefficiencies. So consistent with the above maybe tax:

10% those making over	100,000 baloneys	
20%	200,000	
30%	300,000	Initially up to
40%	400,000	It could go up to
60%	600,000	

And it may be automatically redistributed online on a one to one basis, i.e. by a matching contribution for up to 20,000 to each adult making under 40,000/yr. baloneys.

Because the redistribution is done directly and automatically by the tax code, the government should not need nor should it be allowed to keep more than 10% of the tax revenues for its administrative, organizational, educational and judicial functions and expenses.

The above is an example showing that a much smaller, by over 50% smaller than current governments, very efficient government with high growth potential for the incomes of the poor and with much lower income inequality can be easily achieved.

That is, if the vast majority of politicians weren't mostly pre-occupied about the money they make for themselves and raising money to keep their positions of control over others despite failing consistently in their stated promises and "goals."
Is there anyone who believes that the business people who insist that there is no such thing as a free lunch and calculate return on investment with massive staffs and computers not to miss a decimal, give money to the politicians without expecting and getting big returns on their investments?
Does anyone believe that the politicians are not as beholden to those that fund them as a just paid naked prostitute on their bed?

This is a key reason as to why publicly funded elections are necessary.

Even in the current US system, it can cost very little (in government terms) to fund the very limited in number and frequency Senatorial and Presidential elections and it should be done.

The Federal Election Commission (FEC) may set up plenty publicly televised/streamed discussions, candidate TV/video/radio/internet promotions and debates. All broadcasters may also be required as part of obtaining their license to broadcast at least two FEC organized events every 4 years. Candidates participating in this public financing may raise only up to 20% of their total financing privately. Candidates may choose to not participate in these publicly funded broadcasts and choose private financing.

Further, full public disclosure of all political private financing sources and ads by the name of an individual (and of organization affiliation if any) must be required.

Congress may keep doing their baboon things so long as they pass this Bill for Campaign Finance Reform.

Voting for people who will surely support such Campaign Finance Reform or will at least support 12 year Congressional term limits is central for positive change in the functioning of this government.

In case any Liberal gets new bad ideas, I would strongly advise against high top marginal tax rates in the current militant competitive systems because the rich and their multinationals are less committed to patriotism, other than in talk, than gang banging whores are to any of their customers and will take their money out of that economy as soon as anyone starts even arguing for it in Congress. The rich are the best customers of and for whorehouses and for growing their competitive economies.

In this context however, where the growth of the economy is not the purpose but avoiding the New Independent State becoming a wh.house is, along with growing the incomes of the poor, then keeping the very rich out of that economy by high marginal tax rates, is beneficial to both the current competitive militarized societies and to the New Independent Demilitarized Disarmed Democratic State.

Even though for the most part progress is made by a series of small improvements, at this stage such a change is much more difficult to do in small progressive steps and it is much easier to start from scratch with a New Independent Demilitarized Disarmed Democratic State.

Parenthetically, this kind of change could be called, if one is a techie "leap frogging" and if a biologist a saltation. Saltation means great or sudden change, for which there is plenty of biological evidence that is confirmed and accepted by biologists who do not want to use the term revolution because it contradicts directly Darwin's statement about evolution: "Natural selection acts solely by accumulating slight successive favourable variations, it can produce no great or sudden modification; it can act only by very short steps." (*On the Origin of Species.* p. 471)

There is evidence of both evolutionary, small step, gradual change and of revolutionary, sudden, big change not just in biology, i.e. saltation, but also in nature such as the destruction of the Dinosaurs and as is birth and as is death and in physics such as the Big Bang itself, and in history such as the American Revolution.

For a people who became a nation by a revolution to deny the existence of revolutionary change in the historical record is non-sense.
The correct Theory by which to understand physics, biology, history, people, organizations and life, is the **Theory of Revolution and of Evolution.** (Sorry for the long parenthesis.)

- Hopefully most people will be given enough baloneys so that for their internal interactions and work they don't demand baloneys (a moneyless society,) as is the case in the whorish-beastly hunger games of our systems.

It would be economically very useful not just to these new Independent States but for global improved efficiency if no intellectual right royalties or fees were accepted, in those New Independent Demilitarized Disarmed Demilitarized States.
Or else Greeks may get bad ideas and start charging by each of their letters and words that people use in talking or writing; for example, a Spanish study shows that there about 17,000 Greek words in the Spanish language; and there are over 50,000 Greek words in English. If singers and movies demand intellectual rights for their baloney sentences, shouldn't the inventors of the words they use also have some intellectual rights?

The assumption of scarcity that modern economics is founded on is only partially valid because the primary scarcity is in the spirit, minds and hearts of people— not on material stuff— and it doesn't have to be so.

The most important factor in achieving those objectives is that the citizens of the New Independent Demilitarized Disarmed Democratic State must be mostly immigrants who **freely choose** to not be ruled by force, physical violence, money and appearances.

It would be nice in these Independent Demilitarized, Disarmed (hopefully democratic) States to get rid of all the hundreds of thousands of laws and regulations that we are currently bound by and just keep the "no physical violence against any human for any reason by anyone," the Ten Commandments in your minds and hearts and Jesus' 11th Commandment of "loving one another" (John 13:34) which is the necessary Spirit one needs to be in, to abide by the Ten Commandments.

"**I have not come to abolish the law but to fulfill it,**" says Jesus. (Matthew 5.17)
God's Law was is and will be the Ten Commandments that define the difference between right and wrong; between righteousness and sin, that means separation from God.

-The unjust harm from someone falsely accusing, blaspheming, slandering or from cheating on the agreed rules including in marriage, or from stealing or from killing are well established in almost every legal system in the world and they are the core of civil law, contract law, corporate law and criminal law respectively.

-If you are greedy and desire, envy, covet what is not rightfully yours you lose your wholeness and integrity and if you disrespect your parents your attitude causes unjust harm to you, your parents and to others.

-If you do not rest to restore yourself for at least the Sabbath, one day a week, probably two and for many even three days a week, away from the work of survival to just be and experience existence, be with your family and/or friends, meditate, reflect, and/or pray which is a dialog with God, your ability to be "whole" is significantly hindered.

-Using the name of God to claim superior authority on conflicts among humans that are about the vanity of humans i.e. the political power-economic money- social fame issues, causes unjust harm. For example, see all the, for the vanity of men, religious conflicts. When God is introduced as a justification on these power-resource-ego issues of people's vanity, those who do so should be reminded that they are wrong and as a result they sin.

-God is Spirit, not a physical body and thus should not be depicted as an image. God is the Spirit of Love, expressing His

Love to all life and asking you to live by loving righteously, justly, wisely, mercifully in wholeness, harmony and Liberty.

-God is One and the same one God for all creation. Humans can get to oneness, to being an unbreakable whole; and be one of a kind, unique, only because of and by the Oneness in the One God. Violation of this results in the unjust harm of a person losing their ability to become whole and a unique eternal being.

That is why the simplest and best explanation of the "Instruction Manual" of humans that comes from The Creator, which if you break you get broken, are **the Ten Commandments** that I just briefly explained and which make common sense.

Particularly in the context of the golden rule, which is preached by all seven righteous religions and is valid scientifically because of the reciprocity that exists in Nature through the laws of physics (as proven below) and to which even many atheist agree of "not doing to others what you don't want done to you" the Ten Commandments make common sense.

If you keep those you will not have to learn any of the rest of the thousands of laws and millions of regulations of governments neither by the easy nor by the hard way. The reason why there are so many laws in nations is because some have broken the Ten Commandments so then new laws become necessary about what happens when a particular Commandment is broken.

Relationships, including one's relationship with their conscience and with God, get greatly damaged by the violation of any of the Ten Commandments.

Christ Jesus did not come to provide an excuse for people to keep being wrong and to sinning by violating the Ten Commandments but to teach and get people to believe in right so that they neither want to, nor need to violate the Ten Commandments because they truly and rightly love others.

Love of self is something that we all, and all animals, already have been given and have and it is called the survival instinct. Love of others is what love is truly about.
Because we don't really need to love those whom we like and agree with because we are unlikely to harm them, love is particularly about those with whom we disagree or dislike.

Also "falling in love" is not love; it is a physical attraction that is rooted in common passion, common mutual damage. Converting "being in love" to true love is difficult. That is because you will find that all the traits that attracted you to them become repulsive after a while. So that converting "being in love" to love forces one to learn to at least tolerate everything they thought they hated.
The critical question for this alternative to work is not as much what new cooperative complimentary Independent society should be put in place but whether there are enough women that might be able to love rightly and truly, including physically the poor, the weak and the depressed.

The location of the New Independent Demilitarized Disarmed Democratic State does not matter much, except that it should be small and in a currently mostly uninhabited space, in a rural area, as a "reserve" to preserve human dignity and sanity or in a location that is currently ravaged by conflicts and war and its citizens need peace.

The Independent Demilitarized Disarmed Democratic State should have only a very small part of the nation's territory because until every government in the world is democratic the idea of demilitarizing a whole democratic nation is suicidal. Until all governments are democratic not only defense is necessary but having the strongest defense, for those who want to survive in this competitive world, is best.

The strongest defense doesn't have the purpose of being used but rather it must be least, if ever, used; the purpose of the strongest defense is to deter aggression.

So, until this nation allows the freedom for some of its citizens to live in a **separate** small Independent Demilitarized Disarmed Democratic State, within the territory of the nation, in a "one nation two systems" solution, —that China claims to have with Hong Kong but their interference in Hong Kong's upcoming elections shows that they don't really have— the beleaguered slaves to the beastly, police dominated, terrified, violent hypocritical shameful whorehouses is us.

The other side is that whatever negative names one may want to call, correctly or not, others, because each human has the extreme positive of divine, mostly unused, capacity to reason **rightly**—that even though animals can think and feel no animal has— and the capacity to love others that they may disagree with and/or dislike, all those negatives pale in comparison.

I can understand why the "leaders" who dominate the current hypocritical violent whorehouse systems like them and defend them and being heavily invested in evil they might not want anyone to get out of them. Their deceitful argument is that if their political Party had control of all 3 branches of government they would solve all your problems.

Previous books of this author include THE WORLD ANEW, 2006, that predicted the 2008 crisis and recommended the only solution that would have prevented it i.e. "increasing the capital reserve requirements of banks" as it says on p.386. THE BOOK OF LIFE and THE BOOK OF TRUTH, contain extensive explanations on how to help the lives of most within the current violent hypocritical and very badly run governments of, by and for money.

(I do not say my governments because in protest to the blind guides, I have never voted so that any politician or government that claims to represent me lies.)

-For example, all previous books explain that not imposing a 20% or so import tax or a VAT tax (that the Chinese also have) on Chinese imports to reduce the net transfer of hundreds of thousands of jobs and of $400 billion a year from the US middle class to the billionaires of the dictatorial, oppressive, tyrannical Chinese communist Party, was and is economically, militarily and politically insane, no matter what one's Party affiliation is.

I understand the argument of not taking offensive action against China but one needs not be offensive to the Chinese to require that there have to be measurable improvements in the balance of trade and/or balance of payments deficits.

It is probably preferable to the Chinese to accept import duties by the US to correct the huge continuing trade imbalance with the US rather than pressure to appreciate their currency because even though the surpluses with the US are very high and must get reduced, China's overall global trade surpluses are not that high. (Germany's, by their abuse of the Eurozone, are higher.)

-Not taxing imported, from outside NAFTA, oil or increasing the gasoline tax, by about 25c per gallon now that oil prices are below $60/bl.— primarily because of the Sectarian war within OPEC— to keep demand from growing too much and oil prices down while reducing gas emissions thus taking the best environmental action achievable at this time and reducing fiscal and trade deficits and having money for infrastructure is too dumb, again irrelevant of one's Party.

If this gasoline tax increase doesn't happen and prices fall too far down, below $60/bl. for more than a year, then we are into a "new bubble" and are being set up for the next major oil price shock that will contribute significantly in breaking the global economy in 2016-2017.

Not taking counter-cyclical action as was needed with the capital reserve requirements for Banks in 2006, when the private sector was excessively over-leveraged without understanding the implications, when I suggested it, is foolish.

If Congress and the Administration had any sense, the gasoline taxes would be raised by 25c/gl. now, (they dropped over $1/gl. in two years,) and then if, when prices get to over $130/bl. they would (temporarily) cut those taxes by about 10c/gl., to reduce the damage of getting the economy chocked.
Most of the other suggestions in those books, as those above, are unfortunately, still valid so I will not bother repeating them.
This book focuses not on how to make the hell of the violent hypocritical competitive materialistic systems more tolerable as previous books did but rather on how to get out of them.

I can understand why a materialist who believes in money and gold and does not believe in God, would support these militant competitive systems on the basis that they are necessary and are even the best to increase the physical survival of people, to force reduction of wrong doing and to defend against the greater evil of dictatorial governments.

I do not understand how and why anyone believes that there is something Christian—righteous— about the "leaders" of our governments. Their motives, double talk and actions are very selfish, evil and clearly wrong.
Is it too much to ask that they give the option to some of their slaves to exit their hypocritical violent Whorehouse system?

Because the whorish "rich" are confused, immoral and are motivated by shame and greed; the Media-entertainment "happy hot" stars are profoundly and profanely ignorant, short-sighted and motivated by boredom, emptiness, anger, the need for acceptance and "belonging" in a group: and the "strong" incompetent physical pimps, bullies, beasts in private or of government are driven by fears, let us address some of those fears.

If there is a small demilitarized disarmed, fully Independent New State within the US territory, with the consent and the recognition of the US government, who would invade it?
The two main places in the world that are demilitarized and Independent are the Vatican and Mecca.

Even though the purpose of Mecca and of the Vatican is to preach their religion and is not for common people of any gender and age to live in peace and dignity, as is proposed here, history has shown them not only to be survivable over many centuries but to also be the safest places to be in, while wars and unnecessary deaths are going on all around them.

In fact isn't the survival of both Italy and Saudi Arabia a lot more assured, even in the case a nuclear global war, just because they have the Vatican and Mecca?
The existence of an Independent Demilitarized Disarmed Democratic Holy State within the US would increase survivability, both spiritual and physical, of the US and of humanity even in the worst case scenario.

Are the Italians and the Saudi Arabians the only qualified people and nations in the world to have an Independent Demilitarized State for those within them who want and choose to live by God's principles?

Are Americans too barbarian and evil, even compared to the fn. Italians and the fn. Saudis, to allow one small Independent Demilitarized Disarmed Democratic Holy Land and State within their territory?

One and only one small Independent Demilitarized Disarmed Democratic Holy State of Peace, within almost every nation, for those who are blessed by Jesus is the only right solution, is needed and can be achieved. The blessed are the poor, the weak, the depressed, the disenfranchised, the pacifists, the merciful, the righteous, the pure in heart, the orphaned or abused children and other children whose parents want them to be in peace, and the true believers.

An Independent Demilitarized Disarmed State is a place where love, righteousness, honesty, freedom, joy and mercy are not just hypocritically talked about and preached but are the values by which common people of both genders live and experience.

One cannot use "the end" as means because the end by definition has not been achieved yet so by definition there is some inconsistency between "the end," the objective, and the means i.e. the strategy.
The only means that a right end justifies are the means that are least inconsistent (i.e. most consistent) with that end.

There are significant exceptions to every rule, to every generalization and to every statement. For example, it would be crazy, cruel and absurd to expect Germany to have such a Holy Independent State not even a little village. This is so, because it is unreasonable, utopian and cruel to expect any woman including any German women to sleep with or love any German man if he is not working like a dog or like an ant, is not doing his absolute best to get rich and she is not drunk!

Almost all Nobel Price economists have written, including Dr. Krugman and Dr. J. Stiglitz that Germany's measures on Greece have been wrong. Harvard's Martin Feldstein, has also explained extensively that it is Germany that has caused and is causing the Eurozone debt crisis.

While punishing severely Southern Europe for "being in violation" of the Eurozone Treaty, Germany has been and is in violation of the 1st Article of that Treaty of Maastricht about not having significant interest rate differences between the Eurozone nations.

But Germany faces no repercussions because the "Commissioner" and most Northern European leaders are the heads of the immoral monstrous horrible multi-headed Eurozone Beast on which Satan's blood sucking evil great Whore Germany's A. Merkel sits.

Germany has also been in clear violation of that same Treaty's article of not having unsustainable trade imbalances and with over 6% trade surpluses —which should be no more than 3%—it surely has exceeded those limits as has been pointed out even by the current US Treasury Department, but again faces no repercussions because the leaders of the European Commission are dirty whores of Germany.
The central cause of the economic Depression and conflict not just against Greece and Southern Europe but also against Ukraine and a central cause of the turn against the West of Putin have been and are the mercantilist economic policies of the huge blood sucking Whore A. Merkel.

A German lady has correctly filed charges of crimes against humanity in the ICC against the German government for the deaths, unprecedented suicide rates and humanitarian crisis that it is causing in Greece. I hope there will be **"amicus curiae" briefs filed.**

If the debt of the Greek government is not written off by at least $170 billion —that is about 40% of its debt— Greece must exit the failed, oppressive, abusive and very destructive to Southern Europe, Eurozone and re-establish its own globally attractive currency at fixed exchange rate an SDR that's a mix of the Euro, dollar and Yuan, to avoid a precipitous devaluation; (as China did with the dollar; and now does to avoid higher valuation.)

8. Judgment of very few for the Deliverance of most

"No one is a surer enemy than an ungrateful beneficiary," wrote one of the great Ellins.

History, that the Greeks taught the Barbarians, keeps showing and providing evidence, as shown by the currently intentionally enslaved, intentionally impoverished and defamed Greece that this world is ruled by those who still remain barbarians and are most evil amongst us.

If the Ellin Mystics who go back at least 3200 years, from whom all other Mystics learned, about whom you may have heard through the Mystic, (means Secret) Greek fraternities and sororities of Colleges, knew that their descendants would get punished so much by them bringing Democracy, philosophy, knowledge of history, geography, the arts, architecture, arithmetic, mathematics, geometry, physics, science, medicine, fair competitive athletics, strategy and religions to the Barbarians, they would not have spoken. Who are some of the Ellin/Greeks who invented or discovered or greatly advanced each of these disciplines? Orpheus, Achilles, Odysseus, Homer, Cleisthenes "the Father of Democracy," Pericles of the first Democracy, Heraclitus, Socrates, Plato— R. Emerson said that philosophy are footnotes to Plato— Aristotle, Leonidas, Alexander the Great, Cleopatra, Sappho, Epicurus, Diogenes; Themistocles, Thucydides, Herodotus(historians); Phidias (Architect, sculptor); Sophocles, Aeschylus, Euripides, Aristophanes (Arts); Pythagoras, Euclid, Archimedes (science); Hippocrates (medicine); Emperor Constantine and Helena, his mother, who established Christianity as a global religion, and the Olympians are some of the Ellins (Greeks) who brought civilization to all humanity.

Also, who brought Christianity to the gentiles? Was it the Romans who tormented and murdered Jesus by "the law" and who had no interest in revolting against their own Empire, who brought Christianity to the world? Or was it the Jews who rejected Him and still reject Him?
Or was it the Turks who did all they could including genocides such as of the Armenians and still do all they can to destroy Christianity? Was it the fn. Turks who forcefully occupy still many Greek cities including Constantinople, (Istanbul,) and have turned one the earliest Christian Churches, Agia Sophia that means Holy Wisdom, into a Museum and are hostile to the Christian Patriarchy?

It was the Greeks who were revolting because the last bastion of Ellenic, civilization had gotten enslaved by the Romans in 31 BC. That was because Queen Cleopatra had sex with the wrong man; and fd. with the wrong guy who later ended the Roman Republic also and made it into a Roman Empire; so because he was such a hot big bad a.hole he is now remembered for 31 days a year, every August!

Who did Apostle Paul call "the children of Light"? The first Churches such the Corinthians, Thessalonians, Colossians, Galatians, Philippians, and those of Ephesus were all Greeks who had gotten enslaved by the Romans and were revolting. As were the first martyrs such as Steven and Apostle Philip. They were revolting against the Roman Empire spiritually first, rather than as the Jews chose to revolt by physical violence.
Also the New Testament was written in Ellenic, because of the Greek audience to whom it was addressed and was translated into Latin more than 100 years later.

Was it the French, Germans, Brits, Americans, Nigerians, Russians, Libyans or those of any other nation who brought Christianity to this world and to you? No.
As Jesus Christ prophesied after his meeting with "some Greeks," as is described in John 12:20, it was and it is the Greeks who brought Him, and who bring Him to glory.

And who kept alive other great religions such as Judaism through the Septuagint, and Buddhism through the Bactrian Greeks?
What did the Greeks ask or get for making civil humans out of the rest who were and to some extent still are beastly, mean, self-centered, irrational, violent, whorish Barbarians and whorish violent Babylonians?

After a 400 year enslavement by the Romans, who tortured, Crucified and killed Christ Jesus "by the law," the Christian Greeks, in the East, in Byzantium, revolted for freedom; it was soon after that the Romans started pretending to be "the Jesus people" but the West had been sunk into the Dark Ages which are known for no one having accomplished nothing.

The Christian Greek Byzantine Empire lasted only 1000 years and fell when Constantinople got occupied by the fn. Ottoman Turks in 1450 A.C and became Istanbul. The Ellins who choose to be dead in ego rather than enslaved left the enslaved Greece and like El-Greco and Leonardo got the Renaissance going in the West.

After another 400 years of slavery under the fn. Ottoman Muslim "caliphate" and 2 WW wars against the Germs whom you call Germans and the new enslavement by the Germs and other Eurozone scum, isn't it about time to stop punishing the poor children, through Sisyphus like torture, of those who taught humanity and gave it Democracy and everything else you say that you believe to be valuable?

The current evidence shows that all the Greeks failed because the world, including the West, is intellectually, emotionally and morally, i.e. spiritually, more screwed up than ever. So the Greeks failed to really help you; therefore your leaders don't need to feel obligated by their charity any longer and may stop punishing them now.

O.k., don't tell the Greeks so that they don't feel offended but most are so self-centered that one had to be great to get through to them. They have hugely screwed up and in a sense deserved what they got for tolerating to be getting ruled by 3 amazingly screwed up families for the last 70 years because they prefer people with "name recognition." There was one family of Bitches on the left, Papandreou; one family of Whores, Karamanlis, on the right, and one of Sluts (I am not telling their name because one of them is sort of a friend,) playing both sides.

It was offensive to bad whorehouses to say that they are run like the Greek government!

These 3 ruling families got the country into chaos in the 60's forcing a military coup, and then to bankruptcy in 2009. The bankruptcy was with the "help" of the fn. Germans, and thanks to the New York Federal Reserve Chairman at the time, fn. Tim Geithner, who was advocating and achieving reductions, not increases as he should have, in capital reserve requirements of Banks, thus making sure that the US financial collapse, that triggered the bankruptcy in Greece, happened sooner and bigger rather than later and lesser. Having proven that he can screw things up really badly... T. Geithner got promoted into Treasury Secretary.

Here is quote from Wikipedia about T. Geithner, while he was President of the New York Federal Reserve: "In May 2007, he worked to reduce the capital required to run a bank.[24]"

The hypocrite "regulator" Geithner was promoting less capital reserves for Banks because his "regulated" friends, the big whore Bankers included him in secretive "world rule" groups such as the Bilderberg's and the Trilateral Commission.

It is these "people" who along with other secretive organizations such as the Free Masons and The Business Roundtable, who by funding both political Parties, have been a central part of the rule the world.

It is these "people" who after all their false forecasts and failed predictions, they would go to Church to pray long public prayers for the poor, whose homes they were devouring while getting bailed out for their unbelievably piggish greed and recklessness by taxes that the middle class and the poor had to pay and are paying.

Then in 2014, instead of admitting to having been very destructively wrong the hypocrite Geithner writes a book about how he engineered the saving of the US economy by asking the Banks to "stress test" and see if they had enough capital reserve requirements. Which they didn't; and which he had primarily caused, at least relative to Citi.
So, after it was too late for most and billions of the poorest suffered throughout the world, the bozos did through "Basil 3" what they should have done in 2006 as was advocated in THE WORLD ANEW, (p.386, last paragraph) which is to "increase the capital reserve requirements of Banks."
All the major TV Channels and shows, welcomed, honored and helped cunning T. Geithner —the most directly responsible Federal regulator for the 2008 financial collapse; and most responsible Federal Officer for a recovery that didn't include most— sell his "excellent" book, while completely ignoring THE WORLD ANEW that offered the correct solution to avoid the collapse.

I hope the Federal Reserve Bank and specifically Janet Yellen pays attention this time to avoid the next crisis:
Before interest rates get increased the Federal Reserve should start selling most of the assets it acquired, (approx. $1 trillion,) to finance the economy out of the great recession. Reducing the Fed Balance Sheet by about $1 trillion now that the economy is out of recession is a central difference between monetary policy and the currency manipulation that the Chinese do.
The bozo T. Geithner said that he regrets that "we didn't have the tools" to help the people who were losing their home so they had to help the Banks.
So, you are supposed to believe that the US Treasury, the Federal Reserve and the US government "did not have the tools" to send $14,000 to your bank account.

That is the amount that you would have gotten, as every household in the US would, if the $700 billion "bazooka" money of Bush and the $800 billion "recovery money" of Obama was to be distributed equally to each US household. (Let alone favoring the poorer.)
Would this action have saved the Banks also? Yes.
Did any of that 1.5 trillion dollars end up in your bank account? No.

They increased not decreased the income and wealth gap. It all ended up in Banks and big corporations with "tools" that the Treasury Department had like "Subordinated Notes," "Mezzanine financing with Warrants" because Geithner, H. Paulson and the Federal Reserve "didn't have the tools" to send money into your bank account.

This happens because politicians have been operating like a "den of thieves" helping other "friendly" thieves to control and punish the vast majority of people by thousands of laws, to oppress them financially and work them to oblivion, for their pleasure, and to influence people's "likes" and dislikes through the media.

So they, the blind guides, the den of thieves, and big evil hypocrites at the top, make sure that the rest, their slaves, learn to like getting screwed and suffer every which way but Sunday, both coming and going.

Geithner, also publicly admitted on TV to have approved "with a chill up my spine" "the slaughter of Greece" as (Ignorant Mother Fucking) "European Finance Ministers asked" "to teach the Greeks a lesson and make it an example." This is evidence of an intentional effort by Germany with the consent of the US government to destroy Greece. In 2015,

Greece had to cut further the $500/mo. pensions to the elderly, (descendants of Homer, Plato, Socrates, Pericles and Aristotle,) after 6 years of economic Depression, to pay usury interest to the IMF...

On April 17, 2015, in a Press Conference Obama said regarding Greece: "When the new Prime Minister came in, I called him and I said, we recognize you need to show your people that there's hope and that you can grow, and we will be supportive of some flexibilities in how you move forward so that you can make investments and it's not just squeezing blood from a stone." The implication is that so far it's been "just squeezing blood out of a stone."

Would the US dollar hold up as a common US currency if interest rates in the North were less than half than the interest rates in the South as is the case in the Eurozone?

Can any State compete by borrowing the same currency from the same central Bank at 2-3 times higher interest rates than another State?
Having a common currency while having large interest rate differences as there are within the Eurozone, is blatant huge thievery by the Germans. There should be 0 interest rate differences if there is a common currency.

I have already shown through the Revelation the Devils' troika of deceit, hypocrisy, false "knowing," false prophesy and pretending to be "good" and "the best" "the best of the best" the "very intelligent." It is fear based beastly territorial force and punishments to control others; and Babylonian confusion, shame and greed; and sad entertainment with sex and violence for fun. That is the troika that dominates and enslaves this world.

Breaking up the unholy troikas of Satan particularly of big violent government; big whorish corporations; big dumbing media is the key to defeating evil.

Isn't that why the Mystic (Ellin) Wise American Founders broke the three branches of government, so as to minimize collusion in unholy troikas, amongst the for, by money whores of Congress; the by use of "authorized force" beasts of the Executive; and the looking for the underlying truth in both sides, that sometimes go with the Executive and sometimes with Congress spiritual sluts of the Judiciary?

For reasons explained above, even though Jeb is probably more competent than George W. Burning Bush and Hillary is more moral than Bill's bills, the Parties and voting public that may nominate these families for the Presidency again are very badly screwed up and will cause much suffering.

It would also be cruel and unreasonable to expect any woman no matter how loving to sleep with or love very ugly looking people! For example, unless very ugly looking men like Sen. Reid, who declared publicly that he grew up around 13 whorehouses... but who is counting..., or like Sen. Mitch McConnell have very significant positions of authority and lots of money, it is cruel and unusual punishment for any woman to even consider having sex with them or loving them!

About 90% of the people show the polls, have a poop opinion of Congress.
The 4 legs of the Dirty Pony that poops on your life and being so ugly want nothing to do with each other include the above two ignorant, incompetent and immoral Mother Fucking "Senators." It is to them and the two back legs of the "Elephant-Donkey-Pony" Boehner and Pelosi that you should send links to this book so that they learn to stop pooping at your home and poop on the Chinese and German governments.

Open access to "outsiders" and help for "insiders" who want to exit the New Independent Disarmed States and get back into the "rat race" is critical.
However, it's important that the mostly female leaders of the small New Independent Disarmed Democratic Holy States protect their women and deny citizenship to very ugly looking people like me and those above!

Those who have broken any of the Ten Commandments are sinners. Are sinners "good people" or are they the whores, who cause coveting, work in the darkness and steal professionally following the first profession, look down on you and consider their time more valuable than your; the bitches and pimps who commit violence, bully, boast, threaten and kill in the name of protecting and use the name of God in vain; and the jerks and sluts who covet, lie, pretend, falsely accuse and cheat?

Therefore, doesn't "sinners" mean whores, bitches, sluts, jerks, a.holes, and pimps and shouldn't the meaning of sinner be described by a mostly negative, profane word?

Ap. Paul wrote in Romans 3:23 that "for all (adults) have sinned."
Therefore, each adult has been and someone has in secret in their mind or in honest anger appropriately called them one of the above profanities.

That is the secret that Ellin Mystic Knowers know and can't tell, except to the few select who are not best described by profanity and therefore are not offended by the recognition and confession that they also used to at least in part be so.

Christ Jesus forgave the sinners but He did not forgive the hypocrites.
"Isaiah was right when he prophesied about you hypocrites; as it is written: "'These people honor me with their lips, but their hearts are far from me.'" (Mark 7:6)

Christ Jesus **criticized harshly the hypocrites and warned them of the hell that they are extremely unlikely to escape.**
For example: "But woe to you, scribes and Pharisees, hypocrites, because you shut off the kingdom of heaven from people; for you do not enter in yourselves, nor do you allow those who are entering to go in." (Matthew 23:13)

I was told by a self-righteous Christian hypocrite that Jesus was talking to the specific people of that time and place only and that those statements do not apply to us in this time and place.

That "Christian" obviously missed that the whole and only reason that what Christ Jesus said is so valuable and is right, is that it didn't apply only to that time and circumstance, as what many say may but that it also applies rightly at all times and circumstances, including now.

Isn't the central request of this book at this time in these circumstances to ask the hypocrites at the top who call themselves Christians to not keep inhibiting but to let some who want to enter God's Kingdom do so?

Is the next passage applicable to these days?

"Woe to you, scribes and Pharisees, hypocrites, because you devour widows' houses, and for a pretense (show), you make long prayers; therefore you will receive greater condemnation." (Matthew 23:14.)

How many thousands of houses of widows and poor where devoured during the last financial crisis and how many long public prayers by the hypocrites that were doing the devouring were going on?

"The master of that slave will come on a day when he does not expect him and at an hour which he does not know, and will cut him in pieces and assign him a place with the hypocrites; in that place there will be weeping and gnashing of teeth." Matthew 24:50-52.

The original Ellenic (Greek) New Testament that I read is in my natural language. I use here any English translation of it and do not argue about this translation being better than the other, like the modern scribes, the scholars, do because I consider all translations adequate to get enough understanding and all translations very inadequate to get the full understanding of the meanings of the words, of the names, of the melodies and of the Spirit of the New Testament.

"Woe to you, teachers of the law and Pharisees, you hypocrites! You travel over land and sea to win a single convert, and when you have succeeded, you make them twice as much a child of hell as you are." (Matthew 23.15) (The hypocrites fail to teach the reason/rational and the Spirit of the Law because they themselves do not know it and do not express it by right action.)

Jesus' not forgiveness but chastisement and condemnation of the hypocrites is also in Mathew 6:2; 6:5; 6:16; 15:7; 22:18; 23:16-29; and in Luke 12:56; 13:15. Also, it is in the Old Testament in Psalm 26.4.

Therefore, with the exceptions of the highly invested leaders in the wrongful system, of the Germs of Germany, of those that believe that their forefather was a "selfish germ," of Turkey's Turkeys and of other children of hell, I don't understand how someone can say that they believe in God and in love yet choose to live in hypocritical, deceitful, violent whorehouses and wouldn't want to give anyone else the option of living free, Independently, by the peace, dignity, honestly, love and cooperation that they say that they believe in.

I can understand if some slaves of the wh.houses, like Jenny, believe that there is no chance that they can get out of them. And so they expect God to intervene to liberate them.

Yet, once the Righteous God who reasons and acts rightly intervenes through some messenger and explains rightly how anyone and all those who so choose can exit these deceitful violent, oppressive through money systems, I don't understand why they wouldn't pursue their exodus.

By righteous, I mean the ones who think, reason, and act for the benefit of others, while excluding their self-interest. For something to be called "right," even by the scientific standard, requires the exclusion of one's personal interests and the independent verification of it by others.

Even in politics, Pericles of the First Democracy about 2500 years ago in Athens said: "We alone do good to our neighbors not upon a calculation of interest but by confidence in freedom and in an honest and fearless spirit."

Are any of your current political and corporate leaders righteous? No, they pursue their own self-interest and their Party's interest primarily which is why all these leaders are wrong; very wrong. And they are condemned hypocrites by not admitting that they are wrong.

What does "love your enemies" (Matthew 5:44) mean? Unless your enemies are good, righteous, honest people, your enemies are those who in your mind or privately you describe with a cuss word.

By righteous, therefore I mean those who rightly and honestly love the sluts, whores, bitches, pimps, jerks.... and even the politicians who admit to having been wrong.

"Crisis" in Ellenic, Greek, means judgment.
Hypocrisy that means wrong, selfish and deceitful judgment is the cause of all crises; as the words themselves indicate.

Hypocrites don't explicitly state the criteria for their judgment, under which anyone making the same judgment would arrive at the same conclusion because their real hidden criterion is their self or one's group interest and then they pretend that their judgments or non-judgments are right, i.e. are for the benefit of others, using whatever argument they can find.

Most modern politicians operate as hypocrites most of the time, causing the crises which they then become needed to resolve.

Friends, there is no way an adult can operate in these modern systems without making any judgments.
(When I wake up I need 30 minutes to make the judgment about whether to get out of bed or not!)

There is no way one can make any choice, any decision or non-decision without making an implicit or explicit judgment.

Let us not be hypocrites and pretend that we don't judge others these days, in these systems because we have to make some judgment, explicit or implicit in every interaction and for every choice of non-interaction.

Isn't it time to judge the very few most evil hypocrites? And so deliver from evil all the rest?
To judge one needs a standard; a criterion. Like the meter and the gram that the Greeks came up with, for measuring and thus judging distances and weights.

Therefore, I propose that we define Hitler as a 666 cm. asshole, and compare the rest of the evil by that standard of how much Darkness, how much of a physical or in their soul "dark hole" they express by their words and actions.
Based on the above, here are my general and therefore not applicable to any particular person, unless explicitly mentioned by name, judgments:

CEO's of large public corporations because they are mostly hypocrites, make street pimps and tax collectors look moral.

Big media, because they are mostly hypocrites, make common sluts look like superstars of morality.

Modern politicians, because they are mostly hypocrites, make street prostitutes look very ethical.
Didn't Jesus say that the tax collectors and the prostitutes will enter the kingdom before the hypocrites? (Matthew 21.31.)

How much damage did a physical whore that may have screwed thousands of men do, relative to the damage that for example the Dick Cheney did, in screwing millions in Iraq, twice, and in igniting Islamic terrorism, by leaving bases in Saudi Arabia after the first Gulf war, and so caused the unnecessary deaths of tens of thousands of people?

Despite warnings from people like me who wrote in THE WORLD ANEW in 2006, that 'getting into the Babylon of Iraq is easy; it is getting out of it that is difficult. Getting out of Babylon with your shirt still on is extremely difficult.'

Beyond the immeasurable cost and pain from the loss of thousands of innocent lives, Uncle Sam came out of Iraq's Babylon 1-2 trillion dollars short; in other words Uncle Sam lost his shirt, as predicted; and he is not out, still.

In my judgment the comparison is that of a prostitute's 4cm "dark hole" relatively to fn. Cheney's 222 cm. "bigger than self" dark hole.

The technologically super-advanced "badass meter" I used, has shown clearly that statistically speaking the higher up one is in the political, social and economic hierarchy of their nation, the bigger asshole they are!
There are strong indications that this world isn't run by Dictatorships, Oligarchies—which call themselves Meritocracies, wrongly— and Democracies but rather it is ruled by Kolocracies i.e. Asscracies, in which the bigger assholes rule.

I am writing that the evilness of each person is precisely measurable and it is being measured by God; as is the righteousness of each.

Relative to the hypocrites at the very top of our societies who most selfishly sought power claiming to be right, despite not knowing what they do, for their own ego and caused great damage to many others, the people of the lower upper class, the middle class and particularly of the lower class are saints.

So, you may now rejoice because you have been justified into Paradise if you chose to enter it.

Terrorist, who murder innocent civilians as their means to cause terror, are murderers and should burn in hell. Islamist terrorist who commit murder in the name of Allah, blaspheme God, i.e. cause damage to the name of Elli, and should burn in hell without forgiveness ever. (Matthew 12:31, 32)

The "the ten horns" of the Beast, Satan, ruling the world, who have been condemned and must burn in hell for humanity to be delivered from evil include but are not limited to m.fs.: A. Merkel, C. Lagarde, M. Draghi, (talk about a dragon!), B. Assad, Ayatollah Khamenei, Al-Zawahiri, R. T. Erdogan, putain V. Putin, Kim Jong Un, and Xi Jing Ping.

All the members of the condemned to burn in hell Satan's stinking huge whore A. Merkel's German governments, who are sitting on the multi-headed Eurozone beast, are mass murderers as were their fathers and forefathers and they along with almost the whole adult upper class of Germany, fulfilling C. Marx's wish, should burn in hell unless they change their course against Greece now.

Whoever continues supporting or allowing the horrible huge robbery of the children of Southern Europeans through the Eurozone multi-headed Beast on which the God damned to burn in hell hypocrites sit who talk about "solidarity" but are blood sucking Whores of Angela Merkel and her governments, should be thrown into the Abyss.

I don't understand, whether someone is a materialist or an idealist, if they are poor or disenfranchised or feel that they have no real say in how these systems are run or are depressed or are righteous or are pure at heart or are a pacifist, (believer or non-believer,) why they would want to keep suffering and not exit the system that caused them and keeps causing them suffering.

And I don't understand —whether one is a materialist or an idealist and even if they like the current militarized systems— why one would be so cruel as to not want to allow the freedom for the poor and depressed who feel tormented by these systems to live Independently, free, in Peace, by their religious or ideological beliefs, particularly since this small New Independent State could be of no threat to anyone since it is by definition demilitarized and disarmed.

This is not a political issue, in the sense of it being a "left" Liberal or a "right" Conservative issue; it is not about whether there is a need for a relatively smaller or a bigger government, and where, at this time.

It is about letting those that are very disenfranchised and/or are not benefiting from the current system whether someone from the extreme "religious right" or from the extreme "socialist left," be Independent and thus give them the freedoms that we claim to believe in, so long as they are demilitarized and disarmed.

The proposed New Independent Democratic Demilitarized Disarmed State is intended to appeal to both the religious extreme of the "right" and to the extreme leftists i.e. socialists because even though with a small government, it is focused on reducing most income and wealth inequality while providing most opportunities for the poor.

People from both the extreme "left" and the extreme "right" along with non-voters will be the most likely candidates to exit because they are most disenfranchised, which is why they are at the extremes.

So that allowing people to leave and get into a new Independent demilitarized disarmed democratic State, within your nation, will politically rebalance and moderate both political Parties of the current militarized system, which it desperately needs, realistically.

Since you know that there are millions who suffer greatly within your system and you say that you believe in freedom, why wouldn't you let them have the freedom to live independently particularly if they agree to be non-violent and live without any military and with no arms?

It is probably true that if the poor worked hard to sell more of themselves they would not be poor; so how much should they really get punished for that?

Even if those who want to exit this system are in your view crazy, lazy wrong, ignorant and bad, why would you say that you believe in freedom but do not give them the choice and freedom of an Independent Demilitarized Disarmed Democratic State within your nation? That is condemnable hypocrisy.

Only the ones that hypocritically say that they believe in freedom but in action don't really believe in freedom wouldn't let those that suffer be Independent and thus free.

Both sides benefit greatly from this separation.
The benefits to the liberated are obvious, huge, and I don't need to specify all of them.
The benefits to those who will remain in the current beastly hypocritical money systems are:

From the view point of the rich, "happy" and strong it is the poor, the depressed, the disenfranchised and the weak that have the problem, are the worst performers in these economies, and are the problem.

When a CEO cannot find any better solution to his problems s/he fires some of the worst performers. And it works. Usually the 10% or so of the worst performers, particularly those at the top of the organization, cause over 50% of the problems; so for the most part by firing a few of the worst performers, the performance of the organization improves significantly.

The recognition that all organizational and cultural problems within an organization start from the top made consulting a very lousy business for me to be in. I did it a couple of times and quit it quickly because when CEO's would tell me the problems of their company I had difficult time not saying: I am glad you know how many personal problems you have.

The best leaders in the context of these systems, in my judgment, are those who admit that they were, are and will be wrong to some extent and are focused in finding their errors and correcting them fast.

The poor are getting poorer and the weak weaker and income mobility has been going down in the US (and in most of the world) for over two decades by both the Republican Bush Administrations and by the Democratic Clinton and Obama Administrations and Congresses.

Because most politicians know very little about technology they keep blaming technology, as Sen. C. Schumer did in Nov. 2014, and not themselves, for their increasing failures in what they say but success in getting higher concentration of power and wealth for themselves and their "group."

"Power tends to corrupt," said correctly Lord Acton.

Breaking up of the devilish high concentrations of power and of wealth that for over two decades have kept increasing is what Democracy is about and the people have the responsibility to make sure it happens.

Since you politicians have clearly run out of real solutions for us, the disenfranchised worst performers, why don't you please fire us out of your "beautiful" system? Spit us out of the belly of this Beast.
We are only annoying belly ache to you. Spit us out. It will work for both you and for us.

I am not going to say: "if you like your economy you can keep it" because I don't like getting associated with deceitful people.
Your economy will boom. Government expenses to support all of us bad performers will drop drastically and the deficits will become surpluses. You will be the top global economy again.
Income distribution statistics will automatically improve significantly.
The political system will get stabilized and both political Parties will become more moderate. There will be very few disenfranchised remaining. There will be very few who feel oppressed because there will be a valid, viable alternative for them, to govern themselves as they wish, independently, so long as they are demilitarized and disarmed.

As a result, violence and crime rates even within the current, militarized, systems are very likely to drop drastically. To the extent that this nation is worthy and proves its capability to have a small Independent Demilitarized Disarmed Democratic State of physical non-violence, it will become the realistically very necessary and important evidence and example that non-violence is both achievable and very worthwhile.

This will help significantly further reduce the very high violence within and among the militarized systems, thus producing an upward spiral.

Because no physical violence among humans is God's Will, these Independent, **separated**, Demilitarized, Disarmed, 'Arma get down,' Democratic Lands and States may be called holy.
Your nation and your government could no longer be morally classified as a wh.house because there will be holy people and an Independent Demilitarized Disarmed Democratic holy Land and State in your nation, that is defended by your current militarized government.

Christians have been waiting for the Son of Man to **separate** people "as a shepherd **separates** the sheep from the goats." (Matthew 25. 32)
Neither the sheep nor goats are "bad," even though sheep are better than the more rebellious goats; yet they are both useful so long as they are separated.

This solution is clearly a win-win for both the non-believers, the secular, and for true believers. The benefits from every perspective for both sides, that are being **separated** now, are great and the costs are minimal, if any.

"The person that is True encompasses all and is not partial, the petty person is partial and does not encompass all." (Lu Yun; *The Wisdom of Confucius* Ch.6. 27)

9. The Kingdom of God

"First seek the Kingdom of God and Righteousness."
(Matthew 6:33)

Can you See God's Kingdom or not yet?

The Kingdom of God is the Kingdom of Love.

The Kingdom of Heaven is a spiritual Kingdom in which people live **by Love's expressions through Righteousness;**
Truth, Creation, Oneness;
Enlightenment, Liberty, brotherly love, Joy;
Grace, Beauty, Harmony, Thanksgiving, Kindness;
Holiness, Purity, Clarity, Discerning Right judgment, Peace;
Mercy, Goodness, Forgiveness, Patience, self-control, Charity, humility;
Wisdom, Compassion, Courage, Honesty, Understanding, Knowledge, Integrity, Gentleness;
Affection, Virtue, Fairness, mutual respect, Law, **Justice**, Empathy, Hope, Endurance and Faith.

Christianity; Buddhism; Hinduism; Taoism; Islam; Confucianism; and Judaism, all teach, exactly these Ideas, Ideals, Principles, Universal Values, Self-Evident Eternal Truths and ways of being, each focusing at a particular level, stage, of God's Kingdom of Heaven.

The World Anew, The Book of Life and the Book of Truth each show with extensive quotes from the original texts why these seven great righteous religions of the world indeed teach these Eternal Truths.

Even though each righteous religion does and should teach all of these Eternal Truths, each is focused on a particular level, in the same sequence as they are described; and separated by semi-colons.

The following includes a single quote from each original text of each religion to put some context and to help you confirm on which Eternal Truths each religion is focused.

In the sequence of the religions above:

- "When the Spirit of Truth comes, he will guide you into all the truth; for he will not speak on his own, but will speak whatever he hears, and he will declare to you the things that are to come. He will glorify me, because he will take what is mine and declare it to you. All that the Father has is mine." (John 16: 12-16) The Spirit of **Righteousness**.

- Buddha means Enlightened, by The Light, and is about the joy in being liberated from the life-death cycle and delivered into the "pre-existing" Absolute God. "There is an unborn, neither become nor created nor formed; were it not there would be no deliverance from the formed and created," said Buddha, (UD.80-81.) In the Spirit of **Enlightenment**.

- Krishna has the same meaning as the Greek word 'Christos' from which the word Christ is derived. In Greek it means "Anointed," "Beautiful." In Sanskrit it means "Most Attractive." (Charis in Greek means Grace, Beauty and Gift of Beauty. The word charisma has that etymological source. *Harā that may be written as Chara, in Greek means Joy and in Sanskrit it is the Joyful energy of Krishna.*)

The Lord Krishna said: "At the end of each millennium all material manifestations enter my nature and in the new millennium I create them again; I deliver the holy, destroy the sin of the sinner and establish the righteous." That is the Spirit of **Grace** and that is the Spirit that Hindus and all Mystics are asked to operate by. Mystic is a Greek word meaning Knowing a Secret Knowledge and Knowing God.

- "Righteousness is kindly, and kindness divine and divinity, is the Way that is final." From THE WAY OF LIFE, Lao Tzu, Chapter 16, p.68. In the Spirit of **Holiness**.

- Every Chapter in the Koran starts with: "In the name of Allah the Compassionate, (kind,) the Merciful." This should be a clear clue to Muslims that God is asking them to be forgiving, compassionate, merciful. "Race to forgiveness from your Lord and to a garden wide as the sky and earth, prepared for those who believe in God." (Koran 57.9) In the Spirit of **Mercy**.

- "Wisdom, compassion and courage, are the three universally recognized moral qualities of man," said Confucius, from THE WISDOM OF CONFUCIUS, p.118. In the Spirit of **Wisdom**.

- "Take this book of the law and put it beside the arc of the covenant of the Lord your God; let it remain there as a witness against you. For I know well how rebellious and stubborn you are." (Deut. 31. 24-26) In the Spirit of **Justice**.

Religious leaders disagree about a lot of things amongst each other between religions and within each religion, mostly for self, national and power, authority over others and fame reasons; they operate very different rituals and traditions and will disagree about the priorities among these Ideals, Principles, Universal Values and Eternal Truths.

Even within Christianity, Ap. Peter and Ap. Paul described these Universal Values and Eternal Truths in almost reverse order and priority.

Ap. Paul: "The fruit of the Spirit is love, joy, peace, patience, kindness, goodness, faithfulness, gentleness, self-control." Galatians 5: 22

Ap. Peter: "For this very cause adding on your part all diligence, in your faith supply virtue; and in your virtue knowledge; and in your knowledge self-control; and in your self-control patience; and in your patience godliness; and in your godliness brotherly kindness; and in your brotherly kindness love." (2 Peter 1:5-8)

Ap. Paul: "For we know only in part and we prophesy only in part but when the complete comes the partial will come to an end." (First Corinthians 13: 9-13)

Yet, there is no religious leader or religious scholar, scribe, worth a darn, of any great religion or Sect, who would not agree that each and all of these Ideas, Ideals, Universal Values, Eternal Principles and Eternal Truths, are indeed good, right, true and are part of the Kingdom of Heaven.

Confucius said: "The Absolute Truth is indestructible; being indestructible it is Eternal..."

The Eternal Truths and Universal values are not abstractions but the whole point of each founder of these great righteous religions is that these Eternal Truths were embodied and/or made real by their life, as each human is capable of doing.

Love's, God's expressions are by "the seven spirits of God" (Revelation 4.5.)
The Seven Spirits of the One True Holy Spirit of Love have been shown above and are:

Righteousness, Enlightenment, Grace, Holiness, Mercy, Wisdom, and Justice.

Justice and her expressions are at the bottom of the Spiritual hierarchy (pyramid,) of God's Kingdom — it is called a kingdom because it is hierarchical—because the purpose of each and all of the other Spirits of the Holy Spirit is to avoid resorting to Justice and the Law in which case some not forgiven harm has already been done.

Why does God's Kingdom rule the Heavens?

Because there is no one, not even the worst lawyer being Devil's advocate who can argue that wrong is better than right or that enslavement is better than freedom or that suffering is better than Joy or that meanness is better than goodness or that lies are better than truth because the arguments fall flat on their own face as they do for each of the indestructible, Self- Evident Eternal Truths of God's Good Kingdom of Heaven that should guide the reality of each and all humans.

The Kingdom of God as it is in Heaven, in your spirit— in your intuition, now in your Intellect, in your right reason, also and in your emotions— being also on earth, so that you may experience it from others, faces the obstacle of the hypocritical lies by your oppressive selfish leaders, who are children of hell who actually do all the opposite things of what they know are right and good and make long daily speeches justifying them on the basis that they were necessary for "survival."

Because the evil cannot do what they do in the name of evil, because Satan and his angels have just been thrown down from Heaven per Rev. 12:7-10, they do physical and/or spiritual whoring and violence against other life, deceitfully, hypocritically in the name of your prosperity, protection, survival, religion, right, good and God.

Because the rulers of this world cannot admit that their actions are for self-interest despite harming others, are wrong and are evil, they do their evil actions hypocritically, deceitfully, with lies, in the name of right, necessity by truth, good or religion or God.

The fact that you can read this is evidence that your faith and you have endured the injustices of evil and by your endurance you are now in God's Kingdom spiritually already.

Each religion focuses in teaching a sub-set of the above listed Ideas, Eternal ideals, Principles, Values, Truths, and different Sects within each religion prioritize them differently, each focusing at a different level/ stage of God's Seven Heavens and of the Hierarchy of Eternal Truths.

Truth is and must be hierarchical, with sub-truths adding up into a "greater" truth, to contribute to understanding.

Even physical truths, represented by words, are hierarchical. For example, apples and oranges are both part of the "greater truth," fruits. Carrots and potatoes are vegetables. Fruits and other trees and vegetables are part of ... which along with fish and..., are part of living things, which along with...make up this world...which along with...

Within Christianity the seven Heavens are symbolized — yet not expressed substantively, abused in the past, without correct understanding of their meaning yet and their correct expression in practical life— by the seven Sacraments that are based on recorded actions by Christ Jesus of: **Anointing; Ordination; Confirmation; Confession; Baptism; Communion; and Marriage**.

By understanding that they are in the same sequence as the seven religions listed above and the seven corresponding Spirits of God one may understand the meaning of the seven Sacraments correctly.

The expressions of love through righteousness are necessarily different depending on the needs and purpose of the context.

The **expression of love** in the context of a of a *female-male adult relationship* must be and is different and distinct from the *parent-child love* which is different and distinct from *love in friendship (philia)* which is distinct from *love towards strangers* and different from *brotherly love* that is different from *loving all,* and distinct from worship and *loving God.*

The Ellenic of the New Testament has distinct words for each of these seven distinct, because of the different context, right expressions of love.
They correspond in a bottom-up sequence to each stage of the seven Heavens but unfortunately there are no corresponding English words other than as I describe them by the kind of relationship.
Most Christians have heard of Agape, that is "love of all."

St. John's Revelation starts with what the Holy Spirit says to the seven religions above "the seven Churches that are in Asia." (Revelation 1:4)

If it is not all Greek to you, let those that know the Ellenic names understand which of the seven great religions, "seven lampstands," "seven Churches," the Holy Spirit, the Ellenic "Alpha and the Omega," (Revelation 1:8) guides and admonishes, in each of His messages in Revelation Chapters 2 and 3.

The sequence of the messages to each "Church," in Revelation, in terms of religions, is in the same sequence of the seven great religions as listed.
(Amazingly, there are strong indications that it could be fully valid even the in reverse sequence to the righteous religions listed.)

The messages to the seven Churches also correspond to the seven major denominations within Christianity.

In the context of having understood the sequence, the leaders of the religions should read what the Holy Spirit says to them, now.

The Kingdom of God is the focus of the three Synoptic Gospels of Mark, Matthew and Luke.

The Kingdom of God on earth is an Independent Demilitarized Disarmed Democratic State, where people do not get abused for their kindness and lack of need to defeat or to control others or to have more than others but live Independently, separately, by the spiritual Kingdom of God in reality, on earth.

Isn't that what Jesus Christ prayed for? "May your Kingdom come, may your Will be done on earth as it is in Heaven" (Matthew 6.10)

All seven righteous religions and all Sects within each will benefit and should strongly support having one and only one Independent Demilitarized Disarmed Democratic State within their nation where the Churches, Mosques, Synagogues and Temples may send some of their flock and as a result have as evidence that what they are preaching is not non-existent utopia.

Ultimately each has to make a basic choice and judgment, by which they are judged: Do you prefer the Darkness or the Light?

In particular the "people of the Night" must make the judgment and choice by which they will be judged to hell or not as to whether they will keep enslaving the people of the Light or not.

I cannot understand, unless someone is a child of hell, why one would not want to give the option to their children and grandchildren to live in an Independent Demilitarized Disarmed Democratic State if they choose so and live by Love, Peace and Liberty.

Why wouldn't Christians allow God's Kingdom on earth as it is in Heaven?

These new Independent demilitarized disarmed States are not just an ideological pursuit of righteousness nor are they just a well-reasoned experiment to better balance each nation, in scientific terms to bring it to equilibrium but in some places they are a realistic urgent necessity.

For example, the Taliban are believers in Islam committed to it till death.

There is no way that they will participate as 'happy' members of a secular society, so that destructive trouble, wasted spending and lives lost in Afghanistan will continue, until one small Independent **verifiably** Disarmed Demilitarized State is given to the Taliban and is protected by the UN peacekeepers or by NATO along with the Afghan government.

It could be in all of or part of Kandahar and/or in the Helmond Province and/or in Pakistan's Waziristan; but wherever it is done until the Taliban accept and are given an Independent Demilitarized verifiably Disarmed State, Afghanistan cannot be in Peace.

This is likely to be acceptable to the Taliban because they laid down their arms in 2002 hoping to be left alone, but unfortunately they kept getting attacked.

Another example is that Israel's B. Netanyahu has been calling for a Demilitarized Land in Palestine and France's F.

Holland supports the idea. In my view Palestine needs to be both Demilitarized and Disarmed.

There is an urgent, practical, realistic need for Palestine to become an Independent Demilitarized Disarmed State, under Fatah's Abbas.
Until then the conflict among Israel-Palestinians will not end; there will be temporary cease fires but war will start again and again with, realistically, too many unnecessary deaths and too much destruction.

Right thinking and correct thinking

You currently are required to think about whatever you believe to be for your benefit. Capitalism is explicit about being based in each pursuing their self-interest.
Thinking for self-interest is wrong thinking. It is wrong because it results in the willingness and likelihood to do harm to others for one's own benefit.

As a result of wrongly thinking for self-interest one is willing to sin against others, if they think that they can get away with it. That is why one needs to "deny them-selves" to become a disciple of Jesus, i.e. a Christian.

Denying self—the 'lower,' the ego self— is required by each of the seven righteous religions.
For example, Lao Tzu wrote: "The Wise chooses to be last and so becomes first; denying self he is unified."

Right thinking requires excluding one's own self-interest and thinking for the benefit of others.
It takes some practice to learn to think rightly because our natural instincts and ego are towards thinking for our own self-interest.

As Confucius said: "True personhood consists in realizing your (Higher) True Self and restoring your moral order; and

"The Man who is True is conversant on righteousness, the petty one on profit."

An intermediate step from wrong to right thinking is correct thinking.

Correct thinking takes into account both the key interests of others and one's own longer term interests and produces what people call (compromise) win-win solutions.

Correct thinking is what Democracy that asks people to think for the "common good," is based on.

Capitalism presumes that in pursuing his/her self-interest one is forced to think about and present/sell why "this" is also in the interest of others.

So businesspeople are forced to think correctly in order to succeed despite their wrong intent; but at least they admit their selfish intent; (for money.)

Very wrong thinking is thinking about one's self-interest and then pretending that you have been thinking only for the interest of others which is what the hypocrites and most politicians do and don't even admit their selfish intent; (for power, fame and money.)

Evil operates through very wrong selfish, fearful, greedy and animalistic thinking appearing as if "for your good" but in practice most if not all the benefit is for itself."

All seven righteous religions ask one to intend, to think and **to act rightly** i.e. in the interest of others and not in self-interest.

For example, "He who does his task dedicated by duty without caring for getting the benefit from his work; he is a yogi." (The Gita p. 864)

The central way to avoid hypocrisy is by following St. John's call: "Let us love not in words and speech but in truth and action."(1 John 3.18)

To be a true believer in God and/or to be a true Christian one must start from intending, thinking and acting rightly— i.e. for the benefit of others and provably excluding one's self-interest—based on right understanding of all the critical facts and spiritual truths of the context.

10. Racism

People and nations face economic, defense, political, social, individual, relationship, physical, environmental, moral and religious problems which are interlinked like a Gordian knot so that most solutions to each set of problems aggravate the other problems.

Ethnicity, nationalism and racism underlie and are intertwined with the Arab-Israeli conflict and with the Arab-Iran proxy wars of the Middle East because it is where three of the continents, each of primarily a different race, meet. Racism is the favorite topic of blacks in the US also.

It's natural and not unreasonable for one to like those who are like one-self such as one's family, ethnicity, gender and race.
The problem arises if one dislikes disrespects, demeans and damages those who are not like one's- self, family, ethnicity, gender or race.

The only way to truly overcome that dislike of "the others" and racism, sexism etc. is by learning to think at least correctly and preferably rightly.

Also, any generalization or stereotyping may have some statistical validity—with significant exceptions— and may be useful for understanding but it is invariably prejudicial and wrong in judging any one.

Ethiopia is an exception in Africa, having never been colonized and being mostly Orthodox like, Coptic Christian and having had good relations with both the Greeks and the Jews as recorded in history from the earliest times such as in Homer and in the Old and New Testament.

For example, the Greeks gods would vacation in Ethiopia according to Homer. However, Memnon the king of Ethiopia (at about the time of Moses,) was the nephew of the king of Troy and went with some troops to help defend Troy from the invading Greeks, according to the Iliad.

Eritrea (means red) is a small mostly Muslim country on the Horn of East Africa— neighboring "lovely" Somalia and across the Red Sea from Yemen— that you probably haven't heard of because it was part of Ethiopia when I was born there, (like a 'red heifer,") but finally became Independent in 1993 after many decades of vicious war.

My grandfather was the first industrialist in Ethiopia and a friend of Emperor Menelik; my dad was a friend of Emperor Haile Selassie, who is also called Ras Tafari, so I, despite being white, was born and grew up there, which I liked until we were kicked out (for being white) and had our factories and home and everything else nationalized, without any compensation, still. That was in 1975, by communists led by a condemned mass murderer Mengistu, who also murdered Haile Selassie, whom I briefly met a couple of times and liked.

As a result of my compassion for my common Ethiopian childhood friends, I was a reverse racist and supported the first election of Obama for the most part because he is black, thinking that no one could screw things up more than G. W. Bush.

I even had a blog for a while suggesting weekly meetings with Republicans and emphasizing that the key to Obama's success was the bipartisanship, the "new environment in Washington" that he promised but after his election he hypocritically ignored. It is not in rhetoric that Obama is lacking; it is for the positive bipartisan action and results that we are still waiting. I was wrong.

Obama screwed things up even worse for most people and for the poor, and even for the American blacks. After 6 years in the Presidency one can't keep blaming, with a straight face, Bush for that.

Now I can't be a reverse racist anymore.
In foreign policy it is true that Bush messed up Iraq big time but it is also true that Obama by not leaving a residual force—as was recommended strongly to him by everyone that isn't a bozo— and by his other actions/inactions has messed up not just Iraq but the whole Middle East even worse than Bush.
And Obama has messed up Europe also by his mishandling of Putin.

Now, my primary suggestion to Obama is: Let God's people get out from under this wrongful, police state, beastly hypocritical whorehouse government.

Obama did follow the dreams of his Kenyan father and has been a good President of Africa, (even though he would have never been elected in Africa,) having taken consistently positive actions to promote health and economic growth of Africa— which has been growing at about 6%— with actions such as the Nov. 2014 Africa conference in the US. Unfortunately, the investors were not informed that they may get nationalized, have everything they invest taken without getting any compensation as there are current legal precedents for it.

If you are not truly spiritual— and as a result don't think rightly— and are in practice a materialist then the color of the skin, the size of the nose, the shape of the eyes, how fat you are, the type of genitals and everything else material and its appearance maters and should mater; they are called facts.

Within these militant competitive materialistic systems because racism, nationalism, sexism, age discrimination and discrimination by every aspect of appearance are bound to happen daily by most of every race gender, nationality, ethnicity and age whether upfront or hypocritically, the best that can be done is to keep trying to reduce them by chastising its biggest offenders; but they will all continue being there.

It is likely that there is racism within the police force against blacks. And it must be addressed.

But the police force in the US is way too violent and way too bossy to everyone. They radicalize people by their forceful, bossy behavior. The statistics show it and I have seen them use excessive force, not just to people of color but also to others including to whites, including to me, to women and to teen agers. No one has elected the police to be the bosses of anything.

That violent forcefulness seems very unnecessary particularly since there is much corruption within the police force also.
The Judicial system relies on the police force and so it overlooks many of the problems within it.

The number and percentage in the US of arrests and of sending to jail people which ends up radicalizing them, is the highest in the world and is absurd.
So to get real change, there needs to be a change of how city attorneys evaluate cases, with their record being made public, how they and the police get trained, get evaluated, compensated, reassigned, transferred in a different department of government, promoted and/or let go.

"Executive branch" employees, starting from the Presidents have a cultural problem of thinking that they are the bosses, have gotten intoxicated with power and therefore suffer from pimp personality disorder which psychologists call histrionic personality disorder.
So, they must get trained to be the least bad possible under the tough circumstances they face.

To join the police one must get tested and get some treatment for histrionic personality disorder so that they don't keep acting hysterical and thus not using excessive force, prior to getting or working in any police job.

Racism and particularly racist discrimination among tribes and ethnicities that result in immense damage, destruction, deaths and mass murders was and still is the most significant problem in Ethiopia, in Africa and in the Middle- Near East.

For example, there are 83 distinct languages by 83 different ethnic, tribal groups, with over 200 dialects within Ethiopia alone, making broad education very difficult. Ethiopia is currently dominated by one tribe (Tigre) that has committed atrocities against the tribe of Amharas.

Because I was born and lived a long time in, know and like Africa more than almost all African Americans, I feel comfortable in saying:

Unless you are truly spiritual stop complaining hypocritically about the racism of others because as most statistics also show, it is you that are most obsessed about the physical, about appearances, are most barbaric to each other and to others, making all my insults to the fn. European blind "leaders" look like love talk relative to the cruel, barbaric, murderous acts going on among you. You are most racists and need most to reduce your own very seriously bad fucked up ness.

There is a sure need for cultural change in the inner city black communities led by blacks.

David McClelland's Theory of Needs says that people are driven by 3 basic needs: The need for power; the need for achievement, and the need for affiliation.

So in explicit terms, when the need for power is excessive it results in hysterics (bitches/pimps); excessive need for achievement through money characterizes most narcissists (whores); and the excessive need for affiliation is associated with those with borderline personality disorder, (jerks/sluts).

The financial markets such as the stock and bond markets globally are rarely driven by positive human attributes but rather by negative attributes such as by **fear** of the strong histrionic beasts-bears needing power, stability, control; by **greed** of the rich narcissistic-bulls needing more "achievement" as measured by money; and by **short-sighted** overreactions for quick gratification of happy looking but hurting "inside" attention seekers needing affiliation.

If there is any problem about anything someone is being too immoral and/or too incompetent and/or too ignorant and most often all these 3.

"This troika" is manifest among the characters involved in any problem.

In common folk language so that every adult can understand and relate to it, if there is any problem about anything, anywhere there must be a spiritual or a physical immoral narcissist (whore) and/or an incompetent hysteric (bitch-pimp) and/or an ignorant borderline (slut-jerk) involved, internally and/or externally.

Confusion is central to the babel, in towers of Babel that empower Babylon's whoring.
Clarity is central to overcoming the dark confusion in which the spiritual and the physical whoring thrive.

If you stay in this system new pimping is no longer the "lovely" forcefully enslaving bossing that it used to be; the new advanced pimping is about looking strong and "happy" and talking about non-violence.

It is economic power that has been ruling the world; it is no longer physical strength, force, and physical power.
The fastest way to economic power is to be adaptable and entertaining.
Be loud; and pick a simple beat and put whatever nonsense you say into that same beat and tune. Maybe they will mistake you for Homer or Shakespeare!

If you are black and anyone wants to have a conversation about anything, say that it is all about race discrimination immediately and call them racists!
If the one who wants a conversation is also black, call him by his Greek name, negro, to be sure that he knows that you know what he is!

The police of— a fn. Kissinger supported, CIA installed— military Junta's, a whole bunch of them, kept shooting and chasing me like a dog around the streets of Athens for having participated in the student revolt in the National Polytechnic in which I was studying in November 1973 against the fn. military dictators, for having the gull to ask for freedom of speech in the birthplace of Democracy.

Parenthetically, Geithner was Kissinger's protégé,(worked for him,) and both have long been whores to the Chinese Communist Party and have blocked efforts by both Parties to try to rebalance the (primarily) by Chinese currency manipulation unsustainable trade imbalances.

Within a year, my dad passed away and then the Ethiopian government nationalized everything and made me broke; that was my time of crisis and the source of my passion.

I received a couple of M.Sc., degrees and then graduated from Harvard Business School.
Then, I did business turnarounds as a manager then executive, President, CEO and Chairman of electronic manufacturing companies. So, I became rich, (again) in a young age.

I was acting hot and could feel my soul burning but was the wrong guy from a young age as whichever unfortunate woman was attracted to me found out sadly!
From having narcissistic traits like believing that I am better than others and entitled, to being impatient, inconsiderate, quick to anger, using profanity, selfish, sexist, dismissive, physically unhealthy and with so many more other flaws that I had to use a calculator to count, it was clear to me as a decent, ethical, self-knowing CEO that I had to fire myself away from the nice people whose ignorance I was taking advantage of!

So, I fired myself out this world in 1990, after selling the last electronics manufacturing company that I run, when I was 34 years old, 25 years ago, and have not worked in business since, except spending 2-3 hours every few months rebalancing investments, as shown in the Exhibits.

I was depressed and quit when I realized that there was nothing from all that I knew about being successful in this world that I would be proud to teach my children. What could I teach? This is good pimping?

While in the last electronics manufacturing company I run, which produced an average of 230% return on investment a year to investor-shareholders while I run it, I spent at least a couple of hours a day bypassing my executives and managers, walking around and talking directly to the front line employees, those at the bottom of the organizational hierarchy Charts, who actually do the hardest work, to find out what is really happening and to help if I could find how.

One day someone, wise, who was in Customer Service, asked me if he had any freedom and choice as to what he does.

The honest answer would have been: very limited freedom; so limited that if you continue this, uncomfortable to me, conversation you will get fired. My deceitful answer was... 'yes, plenty of freedom to do whatever you like so long as you keep the customers happy.' Actually, I ended up having the relevant executives promote him, because it is difficult to find people who can or even want to think independently.

Slavery lasts because some like being slaves and others don't admit to being slaves because they keep getting fooled into thinking that they have plenty of freedom.

Why are you offended by "bad" names and are not offended for being enslaved into a violent deceitful hypocritical violent money system?

It is not **being called** a name that you should find offensive and get angry against the messenger but rather it is **being** a slave that you should find offensive and get angry at the deceitful, hypocritical, big bad leaders who enslave you and give you no option to be free, Independent, in peace.

Demand freedom; demand an Independent Demilitarized Disarmed State that isn't a whorehouse.

Freedom requires having adequate alternatives at all levels. Where is the alternative to the single, militant, militarized, capitalistic system that all, in all nations, are enslaved into? You need a choice of two viable systems, two Independent Societies within your nation, both available to any at any time, with free movement between them, both democratic, to be free.

You must give yourself the option of an Independent Demilitarized Disarmed Democratic System, somewhere, to call yourself free. Those who will choose otherwise are (empty) ignorant and/or (angry) incompetent and/or (shameful) immoral slaves.

Friends you have been getting pimped very badly by very few obscenely wealthy obscenely powerful obscenely deceitful hypocrites at the top.

Even though the hypocrite "bosses" talk with language that is nice and proper, give fancy speeches about liberty and freedom, dismiss those that use profanity, their actions are very profane having enslaved you in a spiritual and real Whorehouse and having made you and transforming your holy children into control, fun, fame and money chasing statistics with horrible content of character, as choice-less slaves.

Even though there is a lot of hypocritical talk about freedom, people of all races, genders and ages, particularly the young, talk, sing, dance and act like slaves because they are being made into and feel like enslaved ignorant slaves.

Hypocrites are so bad that they cannot admit it even to themselves and are therefore not fixable.

Get out of pimping, bitching, whoring and being a slut, jerk, as fast as you can. Start running now for practice, go, Go! Revolt; peacefully and non-violently Revolt; Revolt.

Keep peacefully non-violently revolting until you have not only the right to talk freely but more importantly the right to also live freely, independently in non-violence, dignity and peace in an Independent Demilitarized Disarmed Democratic State.

Keep peacefully but actively revolting until you are allowed to exit this enslaving deceitful, hypocritical violent system and into a Free, Independent, Demilitarized Disarmed peaceful State within your country.
Is it clear? Are you crystal clear?

There was a time for rich whores but now "fallen! Fallen is Babylon the great." (Rev. 18:2).

Darwin said that it is not the strongest that survive, nor are the most "intelligent," (i.e. the whores who make the most money) but it is the most adaptable that survive. (Which of these beastly attributes is best for survival actually depends on the time frame that one uses to evaluate survival. As an atheist friend says, morality is the best very long term survival strategy but for these purposes let's use what Darwin said.)

So, according to Darwin, become an adaptable slut to live!

If you are truly spiritual and not a hypocrite in claiming to believe in God, then get out fast from the violent hypocritical deceitful materialistic violent wh.house systems.

If you believe in non-violence or want or practically need to escape the violence exit this system. I showed you how.

If not, and you are a materialist/ realist who will remain within the violent hypocritical corporatist systems, may I suggest that since you are going to be wrong anyway, at least be usefully wrong and consider following a utilitarian philosophy.

For materialists; realists

Friends, Sophocles in Oedipus Rex showed that those leaders, who cannot see spiritually and therefore cannot see that they are ruling violent deceitful whorehouses, are driven by oedipal issues and act like blind mother fuckers, (m.frs.)

Homer called those strong but profane people who get upset at profane language because they are in whorehouses but don't see it because they perceive only appearances and cannot see the content of character because they cannot see depth, as if they have only one eye i.e. Cyclops.

Your leaders have been and are still being called aggressive, argumentative Cyclops and ignorant, blind m.frs.
The Ellins keep acting like they have special connections with the Most High, Elli.
But what and who do these fn. Greeks know?

Prophet Isaiah wrote in 35:5 and Christ Jesus spoke about giving sight to the (spiritually) blind (Matthew 11:5) and warned the (spiritually) "blind guides" (Matthew 23:24, Mathew 15:14) who destructively lead the people to the "pit."

All those claiming to be right or to think or judge rightly or to doing right better prove it with evidence confirmed by others because right is right only if it is proven so by independent objectively verifiable overwhelming evidence.

When this is "that Day" there must be 11 Olympians, Apostles who agree with me and who should organize themselves into the right actions to establish God's Kingdom on earth, as it is being described in this book.

There is scientific evidence that humanity's closest living relatives share about 96% of a human DNA. As usual there is some dispute over the data that puts the range from 86%-99%. It is not unreasonable given that the Genesis account says that God used dust to make man, and presumably the chimpanzees also.
So, that there is at least a 86-99% chance that God used evolution for up to 99% of making Adam. At least 1% of the DNA of humans is unique and revolutionary.

Maybe the different pigmentations of skin among races can be mostly attributed to the long term effects of geography and its climate, its resulting natural environment and the resulting likely exposure to the sun.

If our bodies evolved, then the rational for the necessity for an Independent Demilitarized Disarmed Democratic State can be explained in evolutionary terms as follows:
To evolve from let us say chimp like humanoids to humans, some humanoids, even though they didn't kill their ancestors, had to denounce the ways of the chimps by maybe accepting someone's revolutionary idea of rising through their mind rather than by climbing trees, to become better and to live better; like Adam and Eve.

So, even though it may be that whores are critical to our economy and that bitches and pimps are critical for our protection and defense and that jerks and sluts are central for our entertainment and they may be fine for themselves, one, without hurting them, must move to being beyond them to get to the next evolutionary stage.

This means evolving from being for sure wrong currently into being truly civil and correct if in the current militarized systems and truly righteous if in the New Independent Demilitarized Disarmed Democratic States.

To Atheists

There are 4 main arguments that atheists bring up to argue against the existence of God: The absence of non-kin love; cruelty of Nature; "deep time;" and "not necessary."

-The atheist argument that I find valid, is that there isn't adequate evidence of the existence of right reasoning or of true love among humans; and therefore even if God exists, He isn't manifest among humans.

Further, it is the religious wars among "believers" in God that have caused and cause most of the wars and destruction among humans. I can show exceptions but on the whole I cannot disagree with that argument; and this book has the purpose of changing those facts.

What atheists do that is less wrong than some believers is to rely on their-own conscience and logic to decide what is right, good and true, rather than on somebody's misinterpretation of a religious document. That is much better than using a misinterpreted document, as many "believers" have done and some still do as a hypocritical excuse for abuse, murder and war.

-"Cruelty in Nature" was Darwin's primary argument that goes something like: "What kind of a (All) Good God would allow animals to so cruelly eat each other?"

This argument presupposes that people know what degree of pain animals feel and for how long and why? In my opinion animals in nature experience much less pain than humans because they do not know about the coming soon death; and vegetables experience even less, minimal, pain and there are no data to contradict this.

It also so happens that controlled minimal short-lived pain can be experienced as pleasure as you know from sports and sex.

This atheist argument also presupposes that humans are qualified to judge God, are less cruel and could figure out a better way to Create and make the Universe and to create, revolutionize, evolve and sustain life. All these are wrong, not proven and self-serving suppositions.

A sub-set of that argument was recently made by another Brit, C. Hitchens. "The only difference between Kim Jon Un and God is that one can escape Kim Jon Un with death," he said thus suggesting God is a cruel dictator. Yet, all the evidence suggests that God has never shown up, dictated and forced anyone to do anything, except through Moses telling the Egyptian Pharaoh to let the Jews who so chose, free.

Another version of this argument is "why does God, if He is all Good and All Powerful let so much suffering in the world." God is First All Good and His power is not used to destroy but to Love and Liberate.

As has been explained and foretold, this world is ruled by evil which is why all this suffering exists. God has repeatedly intervened spiritually to convince people to choose and truly live by Good, Right and True rather than by self-serving evil intent, by doing wrong to others and to self and then lying.

It is because people, by their-own choice and contrary to every teaching of God, do what they wrongly think is right for them-selves but is ultimately wrong to them-selves because it is wrong to others and there is reciprocity in Nature, that there is so much suffering in the world.
It is by doing wrong to others that people wrong themselves and suffer.
Even Aristotle, who taught about Democracy and the necessity of high Ethics in it, even though he took a materialistic approach to teach about implementation of the right ideas into correct action, argued in his last (eighth) book of *Physics* that there must have been a First Cause, an eternal First Logic (Logos) and Right Reason, a Prime Mover; God.

If there is no Right there is no point in talking or communicating.
It is by denying the existence of Right, that Satan's "intellectuals" make wrong valid and perpetuate chaotic suffering in this world.

Some hateful of Good and self-centered atheists like R. Dawkins cause others suffering by their own lack of righteousness and by their refusal to accept right reasoning and righteousness.

By God's laws of Nature they will burn in hell. Here's how: The first law of thermodynamics is that energy does not get destroyed, it just changes form. So, the consciousness, soul, character, that makes each one of us uniquely what we are and animates us, does not get destroyed but may change form and specifically because of the second law of thermodynamics will ultimately become heat.

It is true though that no one can escape God's judgment and that Hitchens didn't either. God judges and when one has exhausted all their opportunities to do right and instead of blaming themselves they blame God, then after death they are judged and if one is as sinful as Hitchens, and if they blaspheme God, they are removed from God's Presence and so if they have any consciousness left they experience their bodies burning or being in darkness and eaten by maggots. -"Deep time" in the sense that Creation is not 6000 or so years old but millions of years old for life and billions of years old for things, does not invalidate anything about God, even if correct.

Let us assume that "deep time" is correct.

It may invalidate **an interpretation** of a couple of passages in the Old Testament but nothing in the New Testament.

The Old Testament has two different passages about Creation: one is about Creation, the Intellectual Design, and the second is about the physical making.
There is a big difference between the time to create and the time to then make.
One can create a house, design it, in days but it takes months to make.
So that how these passages are interpreted in terms of the time it takes for each part of Creating is different.

The process of creating and the process of making are reverse processes. To create one starts from the end, from the purpose in the future and through the strategy works backwards to find how to make it happen with what is available now; to make something, as with all practice, one starts with the beginning, with what there is, builds a team of common interests and tries to get to the purpose.

The fact that these are reverse in sequence of time processes has confused many not just about Creating and/vs. making and about Revolution and/vs Evolution but also about Theory and/ vs. practice; strategy and/vs. implementation, Nurture and/vs. Nature, Fiction and/vs. non-fiction, imagination vs. knowledge, idealism and/ vs. materialism all of which are also reverse in terms of sequence of time and operate by opposing processes.

For example, effective strategies start "top down" by the hierarchy; yet efficient implementation starts bottom up.

Light vs. Darkness; light vs. matter; anti-matter vs. matter; soul vs. body; are these the same basic opposition from which all oppositions arise?
What happened to all the anti-matter? It became light; and souls?

The way that I find most helpful to understand these dichotomies is by understanding the difference between the important vs. the necessary, urgent. It is necessary to drink and sometimes urgent to pee yet they are not important; no one will ever be honored by others for having peed.

There are a lot of physical things that are necessary, cannot be ignored but they are far from important. To get to the important one must have taken care of the necessary and have minimized the time it takes one to take care of them. Important is love, is righteousness, all the Eternal Truths and all the first components of the dichotomies listed above.

The first component of each of the dichotomies above is important while the second is necessary.

Both the first component and the second component of these opposing sides happen and have to happen, depending on the conditions each has to deal with, just like the duality of light being both a wave and/or a particle depending on what it is tested for.

God is Love and has made it not necessary for survival for anyone to believe in Him; yet, believing in and practicing Love of others is the most important thing anyone can do.

Even if one interprets the "not deep time" as literally valid, is God limited to one time Creation and to not Re-creating?

Maybe both these two (different in content) passages of the Genesis refer to the new creation by the creation of modern humanity.

Further, we know that time is relative; Einstein proved it. Since when is God supposed to use human measures of time? *Light, at the speed of light, can go around the earth in 24 hours, in one day, as many times as the earth goes around itself in 1761 years.*

So, at time measured by light-time, a thousand (light) years, are hundreds of millions of years in earth time.

Therefore one could get to "deep time" even in the context of interpreting those two passages of Genesis.

Some biologists argue, without evidence, that life on earth started through some germ, bacteria or virus or "gene" X million years ago. Even if true this doesn't invalidate the existence of God either, because God may indeed have used, I believe that there is plenty of evidence that He did, both Revolution and Evolution to make the Universe.

It's a bit upsetting to a Chemical Engineer who understands how the Weak Force operates and the spectrographic half-life measurements of atoms, by which all the "deep time" measurements are done, to have ignorant bozos claim certainty about their measurements, as if there are no conditions under which the half-lives of atoms change. Even though M. Curie, conducted tests to find whether the half-life of atoms changes at very cold conditions and found no significant differences, we know that atoms closer to the Sun for some reason do have significant differences in half-life.

So that the assumption that there are no revolutionary conditions under which the half-life measurements of atoms give highly inaccurate results is not valid. Therefore the certainty that there is "deep time" is not necessarily true.

Even if "deep time" is correct it does not at all invalidate the existence of God, it only further proves the immensity of His Creation, which I Advocate and showed is made through both Revolution and Evolution.

-One of the arguments made above regarding "deep time" i.e. that no matter how likely it is, it isn't necessarily true, is another argument of atheists.

If the atheists could show that the Universe and life can be explained without having to assume that there is God, maybe they would prove the non-existence of God because there is no necessity to believe that there is God. But even that wouldn't prove the non-existence of God.

To start with it cannot be proven. It cannot be proven because as Plank put it: "there is no way that something comes out of nothing."

Yes, there probably is a less than one in trillion chance that out of a random imperfection the Universe created itself. But even then, there is a singularity in the Big Bang, which is another way of saying that the laws of physics weren't applicable and we don't know what happened at that time.

So that even if Theory about the Big Bang is correct it begs the question: and who or what caused the Big Bang? What

145

was there before the Big Bang and what is there beyond the Universe?

Probabilistically the chances of you and I, and the earth existing by random, (accidental) processes and not by Design are less than one in a trillion as admitted by even the most skeptical of the Intelligent Design Theory.

Is it logical to bet on a less than one in a trillion chance that you are random (accidental) and ignore the 99.9999999999% chance that the Cosmos has been Designed by Intelligence— Reasoning, Logic— by what the Ellins called Logos?

"Because there is Logos (Logic) in the Universe that is why we call the Universe Cosmos and not Chaos," said Plato.

Isn't betting one's life on a less than a trillion probability rather than on the overwhelmingly likely and rational probability very stupid?

This atheist argument is more absurd than me insisting that your house and car were randomly self- assembled over billions of years and were not designed by anyone.

What are the chances that your house and car haven't been designed and built by someone recently but rather randomly assembled themselves over billions of years; even if I carbon or uranium date some parts of your house and car to billions of years? They are practically zero.

The author of a recent book "The Accidental Universe" Alan Lightman argues that the Universe is accidental and was caused by the second law of thermodynamics that makes improbable things probable. In a TV show when asked why there is a second law of thermodynamics he said that he had no good answer for it other than "for the same reason that 2+2=4."

To start with Mr. Lightman is scientifically wrong because the second law of thermodynamics could not have operated before Creation, causing creation, because the law operates through time on matter and heat, which occur only after the Big Bang.

Second, the reason that there is the second law of thermodynamics and that 2+2=4 is because there was and there is logic, a universal logic that governs things, as evidenced by the facts, whether one believes in logic or not. There is that less than a trillion probability that the Universe is accidental and if one calculates in more detail all the coincidences that had to take place for us to be, it may turn out to be much less than one in a trillion and closer to Plank's constant. It is that probability and the resulting uncertainty that allows for the freedom of the living creation; and it also allows for the intervention of God in the form of coincidence.

The existence of Logic governing the Universe is self-evident intellectually because logic is what the intellect and intelligence is.

There has also been physical evidence of the existence of Right Logic, Logos/God also, embodied in Christ Jesus, so that even the least willing to believe could believe.

The self-contradiction of the nihilists and the "accidentalists" is obvious: They argue by some logic, that they claim is logical and therefore that you should accept, that there is no logic.

If they believe what they claim and there is no logic they shouldn't be talking; because communication itself is an expression of some logic.

Would you believe that your house or car was an accidental creation over billions of years even if the scientists insisted that it is possible because the second law of thermodynamics makes improbable things probable? No.

You would call me crazy if I insisted that your house or car was not designed, because you can see the design, its logic, its purpose and function.

Given that the design of the solar system, of the earth, of plants, animals and humans is clearly superior to the design of your car and of your house, how much more crazy is the argument of those who claim that the Universe was accidental?

If a logical Theory that explains everything in the Universe is found and proven logically, without necessarily assuming the existence or intervention of God, wouldn't that just further prove that there is Logic and Intelligence in the Universe and that therefore there must have been a rightly reasoning, logical God, Spirit, who created it, as Einstein also believed?

Do the laws and logic of deterministic cause-effect that govern gravity and the logical principles and probabilistic mathematic laws of what I call effect-cause i.e. purpose to cause, of symmetry that is the quantum equivalent of reciprocity, of duality and of synchronicity that govern the 3 quantum forces, invalidate the existence of God or do they prove that there is Logic by which the laws that govern the Universe are operated?

The scientific findings prove the existence of reciprocity as a law of Nature, proving the "golden rule" of the prophets.

They prove, not disprove, the existence of an Intelligent Logical God, —Logos— whom we know.

It seems to me that the cause-effect law of physical solids, i.e. the action-reaction law as per Newton, is the Karma; and the desired effect i.e. objectives to cause, quantum principle is the Dharma of Buddhism. Dharma is about "doing what is right," doing what is for the benefit of others.
Doing right is also a basic rule of I. Kant's (otherwise mostly wrong) philosophy.

As A. Einstein and as most great scientists believed, the existence of logical laws governing the Universe proves rather than disproves the existence of an Intelligent, Logical God.

Because of the spiritual and physical reciprocity in Nature a wrong action produces a wrong counteraction and an increasing vicious cycle; and vice versa a righteous action produces an upward virtuous cycle.
That is why right action ends up producing the right results; proving that it is right.

Also, the "infinite energy" released by the "singularity" of the Big Bang proves rather than disproves the existence of an all-powerful God.
Egocentric scientists using their logic and/or the logic they discover in the laws governing the Universe to argue that there is no logical, Intelligent God is self-contradicting non-sense.

If one is a materialist, realist, pragmatist, then the only way to go is to look at the facts, scientifically and then make reasonable hypothesis about why things work as they do, then experiment to test and verify objectively one's hypothesis, theory.

The job of science is to develop a logical Theory about what are processes and mathematical laws that govern the Universe and Life and to prove it.

Any theory —including the "string" and super-symmetry Theories as currently constructed— that cannot possibly be proven is unscientific.

Because there is no way that I know to prove a negative and because for sure there is no way to ever prove that there was, there is or there will be no God anywhere in the Universe or beyond **atheism is unscientific**.

Let alone that atheism is too stupid, with much less chance of it being true, than me being Zeus, the Pope, Mahdi and the 12th Imam.

Do you know the Mystic, the 12th Commandment?
The 12th Commandment is: **Though shalt not be too stupid.**
This is so because not even God can save someone who is excelling in stupidity.
Can someone who insists in banging his head against the wall be saved?

Being Agnostic, that means "not knowing," is not inconsistent with science and it is admirable honesty, however being ignorant is nothing to brag about or to call for followers.

-Apply, if you like at paulzecos@gmail.com for your not-slave but Master Degree Certificate that you earned by reading this for only a million dollars. If you don't think that is an amazing once in a lifetime time deal... then screw you!

-For people for whom money is an object, there is the Bachelor of Philosophy, Science and Religion degree that you also earned, if you read this book and answer 1-10 questions, with over 80% correct answers, for only $1000, which is over 90% cheaper than almost any College.

-Or you may get the Doctor of Philosophy, Ph.D. digital certificate, from the new proudly unaccredited Universal University, that you also earned by reading this book and writing a review that says that this book is either excellent or the best new book or unbelievably outstanding or shockingly true and useful or any other review with 5 fn. stars, for a voluntary donation of only $12.

-And/or you may say, do and live by what is right and prove it, which includes acting to establish one small Independent

Demilitarized Disarmed Democratic State within your nation and get your Olympian Degree certificate for free and honor forever.

Of course the Universal University is willing, eager and happy to compete with any College in terms of any ethical test that evaluates any of their, vs. our graduates on basic knowledge and understanding in each and all the above 12 Academic Disciplines.

The Sword

"I have not come to bring peace, but a sword," says Jesus in Matthew 10.34.

Right judgment is the double edged sword that breaks up the rule of evil in this world.

Right judgment is the sword that divides and separates the righteous from those who are wrong, the honest from the hypocrites and the Resurrected New from the Old.

"One nation under God" is currently in name and paper only. In reality, the unity of each nation has more holes in it than Swiss cheese. People are more divided from each other and from their government Institutions than ever.

To get to true Oneness and Unity of this nation under God there is the necessity for this separation that gets the righteous, the poor, the unhappy and the weak into an Independent Demilitarized Disarmed Democratic State within the current territory of this nation.

The Independent Demilitarized Disarmed State, within a State, is consistent with the "cities of refuge" that God called for in the Old Testament, in Numbers 35:15, in Joshua 20, and in Deuteronomy and along with Israel are the Promised

Lands that are referenced over 170 times in the Old Testament.

Eastern Ukraine is another case where a verifiably Independent Demilitarized Disarmed State maybe the only solution that is acceptable to both East and West and to stop the conflict with and by Putin.

The need for security in the borders, of a "buffer zone" around the borders of paranoid national leaders is the essence of the conflict in Eastern Ukraine for Russia, as it is the same in essence of N. Korea for China, as it is of Kashmir for India, as it is ISIS for Turkey, Palestine for Israel, and as it is Yemen for Saudi Arabia.

Only Independent Demilitarized Disarmed nations in those "border areas" will provide a real permanent mutually beneficial end to those conflicts.

Still, the valid argument of atheists about the lack of right reasoning and of true love among humans and the hypocrisy of "believers" who talk about these but by their actions manifest the worst cruelty, meanness and irrationality, must be addressed.

The most glaring current example of this hypocritical cruelty and hate in the name of religion and God is Islamist terrorism.

My understanding of the central difference between Islamic and Islamist is that Islam is a religion based on the Koran, while Islamists view Islam as a political system also like that of the 7th century AC.

11. Defeating Islamist Terrorism

The Middle and Near East have been at war since the earliest known history of humanity. Since before the Exodus and after that, during the Trojan War that Homer described, there has been war after war there.

Did Moses, King David or King Solomon and their wars solve the problem? Did the brave Achilles and king Odysseus' wars? Did the wars of the Assyrian, Babylonian or the Persian Empires solve the problem? Did the Ellin Greek Macedonian Alexander the Great's wars solve the problem? Did the Roman Empire wars? Did the Greek Byzantine Empire's wars? Did the Ottoman Empire wars? Did the British Empire's wars? Did the last 2 German world wars? Did the last 2 US Bush wars in Babylonian Iraq solve the problem?
The hypocrite G. W. Bush in his first speech announcing the war in Iraq said that we are going there to "disarm Iraq" and then spent 6 years arming Iraq and training them with weapons, that now ISIL is using.

Now, because of the "brilliant" Obama strategy of not knowing what he does yet being "fired" up and ready to go without getting anywhere, saying baloney in good style, as is popular with 'oi poloi," there is a third war, within a generation, of the US in Iraq and now even in Syria against a Sunni Islamist terrorist enemy with a different name, ISIL, (which being grammatically challenged I spell ASSILL.)

Ideologically, Islamists believe that the Western way of life is not for them and that the cause of their oppression and the biggest obstacle for them living by what they hope is a better way is the West.
Therefore until the West is seen as part of the solution for Islamists to try to live in peace by their faith in some

Independent Sunni State, too many Islamic extremists will keep targeting the West.

Pragmatically, there is no way that the Middle East will be in lasting peace no matter how the war against ISIL is fought and/or won, unless the ISIL held territories in Northern mostly Sunni and of other minorities Syria (and/or of Western mostly Sunni, Christian and other minorities Iraq) is made into an Independent Demilitarized Disarmed State, defended by a small, 10,000 or so UN Security force or by a broad coalition force mostly from Sunni Muslim nations with particular effort by Obama to include Indonesia, Egypt and Pakistan.
This "safe zone" can be achieved only by Obama co-operating with Putin and Assad (and even the consent of ISIL, should be sought.)

The power/money, ethnicity and racism part of the causes of wars have been discussed. Religious conflict among Sunnis and Shiites (which is actually a political/ethnic conflict that started with whether the relatives of the first vs. the last wife of Mohammed are the 'legitimate" heirs of Mohammed) as well as other religious conflicts among people who are utterly clueless about who is God and what is His Will has been a major underlying cause of wars in this world.

Do you Know God and His Will?
Allah is derived from Al-Elli and means: "the God." Do any of the Muslims know the God, Elli?
Elli, pronounced in Hebrew as something like "Ehlee," is also who Jesus referred to in Matthew 27.46 and in Mark 15:34. Islam means the surrender of one's will to God's Will for Peace.
In the **Qur'an** Jesus is referred to many times and is given the title of "al-Masihu **Isa**," The **Messiah** Jesus (see Surahs 4:157,171; 3:45). Isa is derived from the Greek Ἰησοῦς, as is Jesus. The quotation mark like, sign in front of the I, is called

Dasia, and makes the vowel, I in this case, and E in the case of 'Ellenic, low octave, "bass," deep in tone, low pitch. There is no one else other than Jesus, Isa that is referred to in the Koran as The Messiah. Messiah means Anointed, Savior and Deliverer.

The Muslims who can read the Koran should know that the only Savior and Deliverer is Christ, The Messiah, Jesus. No one can be saved and delivered except through Christ Jesus. The reason is that no one can be saved without accepting at least as a matter of faith if not in practice yet, right reasoning and right action. If the Koran is understood and interpreted correctly Islam is no more or less than another currently very self-divided very large Christian denomination. Ironically, the Muslims have been waiting for a Mystic Ellin hiding in his Jacuzzi-well to tell them what they are talking about.

Unless the vast majority of Islamists and Muslims are huge hypocrites why would they insist in solutions for Syria and for Iraq that are preparations for the next war and not for an Independent demilitarized disarmed State for the currently terrorist ISIL held Sunni territory that is the only right solution of true and lasting peace?

There is no rational reason why Russia and China, who also face serious internal threats from Islamist terrorism, will not join in the agreement for an Independent Demilitarized and Disarmed State in the currently ISIL held territory.

An Independent Demilitarized Disarmed State for the Sunnis, refugees and religious minorities of Syria (and/or of Iraq) is the only solution that Russia and China would agree with the West about because it does not disadvantage them militarily.

Both the Obama strategy of Assad must go now and Putin's strategy of supporting Assad are doubling down on failed strategies.

The destruction of Syria and the refugee problem and the ISIL problem will not end until Obama cooperates fully with Putin on Syria.

Only by an agreement of the UN Security Council to establish as a "safe zone" an Independent Demilitarized Disarmed (mostly Sunni) State, on the currently ISIL held territories either in Northern Syria or in Western Iraq or both, will the military defeat of ISIL be achieved any time soon—and it needs to be soon to avoid getting an attack in the US— without much unnecessary bloodshed because it will cause the most defections from the Sunnis in the ISIL held territory.

To enforce the UN resolution a broad military coalition to include Russia, China, Egypt, and with the agreement of Assad and Iraq and other Muslim nations, along with some US ground troops should take over and Demilitarize and Disarm the ISIL held territory.

ISIL is a Sunni movement and is supported by some Sunnis and maybe some Sunni governments. The actions show that while most likely governments to provide such support to ISIL such as of the Gulf States and Jordan have shown that they do not support ISIL by them bombing ISIL the only Sunni government in the region that hasn't done so is Turkey.

Turkey is by action and inaction the national government most supportive of ISIL. The snake Erdogan supports ISIL as a "buffer zone" from Iran's influence.

It would be therefore very wise to move some of the bases from Germany and Turkey to Greece (whose economy is also bearing disproportionate burden from refugees) coughing up to Greece a couple of billion dollars/yr., rather than to the Germs and to the Turkeys.

Once and only once a demilitarized disarmed Independent State is established in these ISIS held lands will Sunni Islamist terrorism be defeated in terms of losing its weapon means for terror and its territory but most importantly losing its ideological cause for terror.

Isn't this right answer so obvious that even those who love money with the hope that some "classier" physical whore might love them for it, can see? Isn't the right answer so obvious that even your political leaders, these hugely bad beastly hypocritical pimps that say: we believe in non-violence and in Peace...unless we feel scared...can understand?

Therefore, while this cohesive strategy by our governments of combating Sunni Islamic extremist terrorism, spiritually, morally, religiously, scripturally, intellectually, intelligence, media, financially and on humanitarian basis is necessary along with a very active cohesive ground objectives and military strategy as to what and how the currently ISIL held territory will be taken over and governed and by whom, as above, the best response from the prospective of a citizen to any form of terrorism is being unafraid and stoic to mitigate the drama of the hell that Sunni Islamist terrorists are in and are trying to get us into.

If this change to establish one small Independent Demilitarized Disarmed State that is not ruled by whorish wrong hypocritical judgments, by prejudice which is the result of hypocrisy and by persecutions about "the law" but rather by honest, righteous, merciful love has not happened yet, despite great men's teachings, revolutions and religions, why would it happen now?

Now, because of communications technology the (world) public has broad fast access to read and to learn fast about what they do.

"You who believe be maintainers of Justice, bearers of truth, for God's sake, though it may be against yourselves, parents or relatives." (Koran 4.135)

Even though the methods of Islamists are extremely wrong and are condemned there is validity in the cause for rebellion against the unjust great suffering experienced by the vast majority of people in the world and particularly of the Middle East.

This unjust great suffering of billions of people is causing righteous anger, wrath in God. God's wrath is the one expression of anger that for sure-sure one should not experience. Change; that is what repentance is.

Deny your lower self, the ego, which is about fun, sex, fame; control, and money; and accept your Higher True-Self that is about right and truly good.
Not in words because the nice hypocritical words have been said all along; change in action. Take action that is truly right for someone else.
Force a change in how your leaders operate and what they do. You can.

Let me make it clear:
Until there is at least one small Independent Demilitarized Disarmed State in the Middle East, the Middle East and as a result indirectly or directly the whole world will remain at war, with increasing suffering, increasing poverty, increasing terror, increasing deceit, piggish greed, immorality, anxiety, despair, anger, physical diseases and pains, increasing percentage of people with serious character disorders and increasing unnecessary unjust deaths of innocent people. Change rightly and demand right change.

12. Jews, Christians and Muslims

I don't really feel like dealing with the issue of the Jews and Palestinians. I am neither and don't care. They have been at war with each other for over 3000 years that we know of. But the Israelis and Arabs are not just continuing their killing sprees but are drawing all humanity into them.
Also, the Jew Jesus Christ who sent me and keeps talking to me is not letting me ignore the issue.
So I will waste my time as everyone else did before me in talking to these hateful, deceitful, irrational hypocrites.

The Jews believe, they say, in God even though they don't truly Know Him; that's why they're not allowed to even pronounce His name.

So let us look at the arguments from this last Palestinian-Israeli conflict.

"The difference between us and the Palestinians is that we avoid civilian casualties and they pursue civilian casualties," said Netanyahu.

The words of hypocrites contradict their actions.

So let us see if what Bibi says is consistent with the facts: 'Because you killed ten of us intentionally and we are better because we avoid civilian casualties we have killed over 1000 of your civilians including children, which we regret but have done nothing to restore.' That is what they did. How is this self-consistent or consistent with a nation that says that it is founded on God's Ten Commandments which without any ambiguity say: "Do not kill?"

"Allah, is The One God, our God and your God, is merciful, compassionate and great; so we will kill you," is the self-contradicted position of hyper hypocritical Hamas.

Who by the way, along with the Israelis, add hhhhhh's to everything so that you feel that they will spit on you at any time, and they do; so it is best to stay at least...one mile away, when any of them speak!
Loving rightly while not becoming a part of the problem requires keeping the right distances.

The central issue in the Palestinian-Israeli problem is not: security; borders, Jerusalem, and the refugees, as they both hypocritically say.

Those are easily solvable issues and most of the world already knows the answers.
The central issue is their hyper hypocrisy that makes every day hypocrites look virginal. These people don't even wait 5 seconds to contradict themselves with their actions and contradict on the largest scale possible.

'We are merciful by faith; so we hate you; die by our actions,' is the official Palestinian position.
'We are Just by faith; so we'll treat you unjustly, as slaves, in practice,' is the official Israeli position.

The Israelis claim that they must be treated as the best allies of the US, despite dissing the US regularly, because they got a clue and picked up the concept of Democracy 2500 years after it was invented and practiced effectively in Greece.

The God of the Jews, Elli, has made it perfectly clear to them that He is not one of them.

The God of the Jews, the God of the Old Testament, has made it repeatedly clear to the Jews that He expects them to keep their Covenant of the Ten Commandments with Him, for Him to keep His Covenant with them. That is what Covenant means; mutual agreement.
The God of the Jews, who got them out of slavery from Egypt by the Exodus, has made it clear to them that they better fear God.

"The fear of the Lord is the beginning of Wisdom." (Proverbs 1:7 and in Proverbs 9:10 and in Psalm 111:10.)

The Jews instinctively fear God, Elli, because He has cut some of their penis off already. They know who has cut off some of their penis already and even the consideration of these rebellious people to revolt against their God causes them to instinctively cover their genitals.

The Jews better fear God and fully abide by the Ten Commandments.

If they fail to keep the Ten Commandments, His Covenant, they will be justly punished by His Laws embedded in the Universe and in Nature.

Are the Ten Commandments too complicated to understand? Is "do not kill" too complex to understand? Is "do not covet" too complex of an idea? Is "do not cheat" confusing? Is "do not steal" not clear enough?

Yet the Jews, have been failing to keep God's Commands and they have been getting punished, as their history shows.

The Old Testament is full of examples of the Jews being hypocrites; saying that they believe in God and that they are fighting for the True God, while sinning, violating and cheating on their Covenant with God.

Moses him-self, and Isaiah and the rest of the Old Testament testifies against the Jews, as shown above and:

"They rejected His statutes and His covenant which He made with their fathers and His warnings with which He warned them. And they followed vanity and became vain, and went after the nations which surrounded them, concerning which the LORD had commanded them not to do like them. 16They forsook all the commandments of the LORD their God and made for themselves molten images, even two calves, and made an Asherah and worshiped all the host of heaven and served Baal. 17Then they made their sons and their daughters pass through the fire, and practiced divination and enchantments, and sold themselves to do evil in the sight of the LORD, provoking Him." (2 Kings 17:15-17)

Have they learned from their history, do they still remember their history in exile and have they learned from their prophets? No. Instead they convicted, and still deny, one of their own, a Holy wonderful Jew, the Son of God, the Most Beloved of God.

In the Israel-Arab-Iran conflict and in the Jews-Muslims conflict I have to favor the Jews but not because of anything the Jews did, but only because of one Jew, Jesus, whom I have chosen as my Lord because He is Right.

If we take Jesus out of the equation for a while and evaluate Jews on their own merit, it is very questionable whether they or the Muslims are the biggest hypocrites.

Did they learn since then? No. Did they learn from the Holocaust? No.
Who is it that has been pursuing money with more passion than any physical whore and is trying through their money and AIPAC to dominate US politics following their bad habit of doing so in the other people's countries including in Egypt in the very old days and then in Europe and then and in Germany before WW2?

Are the Jews not trying to dominate through money, working tightly amongst them-selves, Wall Street, the Banking system, big public corporations, the legal system, Academia and Health Care in the US still today?

Are they not operating as the biggest whores, chasing still frantically money and gold and making idols of gold and other idols?
And how much influence do they have in Big Media, in Hollywood and in the entertainment industry?
Jews conspiratorially amongst each other, go after Satan's whole troika of most money; most control; and most entertainment.

The Jews are worthy intellectual competitors but must stop thinking that they are the smartest because they are not. I didn't have much personal experience with Jews until I went to New York, at Columbia University. There was a bunch, like 7-8 Jews and a couple from Israel in my class who were friends and studied together and were regularly bragging in classes that they are the best and will rank first, and would argue as to who amongst them will be 1st, 2nd, 3rd, and so on.

Unlike what may appear from these writings I do not use profane language in talking and actually rarely talk or socialize because I find it too much effort for too little reward. There is no way to explain all the complexity in life just by talking without writing it and even having diagrammatic Exhibits. I would rather be quiet in a quiet place alone and listen to God. But I ranked first in the M.Sc. Class of Industrial Engineering in 1978 at Columbia University mostly to look at the rearrangement of the dynamics of the neuron synapses in their brain and the resulting facial expressions of those Jews watching the publicly posted results and confronting the reality of having been found clueless!

And who is it that keeps risking the destruction of the whole world, with a potential global war, for some ridiculous housing projects?

Did the True God, Elli, of the Old Testament give the Jews the impression that He can be fooled or that He does not mean what He Commands?

As to the Palestinians, they seem to be descendants of the Philistines of the Old Testament. Philistines, is a Greek word, and my understanding of its meaning is: "love being sick."

The current leaders of the Jews and of the Palestinians make physical gang banging whores look morally excellent!
The current leaders of the Jews and of the Palestinians are huge hypocrites doing all they can for Satan, not for God.

The message that I have been sent to deliver to the Jews is a warning. Either accept Jesus and get saved by one of your own, a truly wonderful Jew, or keep the Ten Commandments without any transgressions or experience the wrath of God and burn in hell.

No iron dome can save you; only a fn. Greek can save you. If you are waiting for an Ellin to save you despite you not keeping your part of the agreement, it isn't going to happen.

It was our fathers that fought effectively the Axis of Evil of Mussolini's Fascist Italians, Hitler's Nazi Germans and Turkey's Islamist turkeys, with over 20:1 population disadvantage, before and after anyone else got into the WWII. The Greeks protected the Jews after they were occupied, refused to disclose who were Jews amongst them and many were murdered for it.
For example: "The 275 Jews of the island of Zakynthos, however, survived the Holocaust. When the island's mayor, Lucas Karrer (Λουκάς Καρρέρ), was presented with the German order to hand over a list of Jews, Orthodox Christian Bishop Chrysostomos returned to the amazed Germans with a list of two names; his and the mayor's. Moreover, the Bishop wrote a letter to Hitler himself stating that the Jews of the island were under his supervision.[17] In the meantime the island's population hid every member of the Jewish community."

Greece was demolished, ravaged, pillaged and mass murdered by the Germans without any compensation after the war for the about .5 trillion euros in war damages, as recognized by some even German Professors.
Greece didn't even get any of the help after the war that even the nations of the Axis of evil got.
No, they were punished and are still punished, in practice, for defending right, despite the nice words such as those of W. Churchill who said that "heroes fight like Greeks."

Contact Merkel, Draghi, Obama, Barroso, Laggard and urge them to forgive at least $170 billion of the Greek government's debts as they should have done long ago because if you don't and if they don't, there will be a time that you and/or your government and/or your children will desperately need the forgiveness of your spiritual and/or financial debts and they will not be forgiven.

It is particularly stupid for the nation with by far the largest debt, the US, and for the Western nations that mostly have very high debt to GDP ratios to not make sure that there is solid precedent for creditors (for the Chinese government) for some forgiveness of debt, through Greece.

Enslavement into the Babylonian rule of money happens though debt. People got and get enslaved into work to pay their financial debts. The debt-holders are ultimately the 1% of 1% of 1% who enslave by debt all others. All others are enslaved by debt, by mortgages and bills they have to pay. The spiritual debt of the few financial debt holders is about as much as their financial debt (bond) on all others.

A necessary change to this debt based worldly system is to establish a law —and an international law by the UN for allowing controlled bankruptcy of nations also— that allows anyone, any corporation and any government that shows that it is in financial difficulty, before having to declare bankruptcy and seek protection from its creditors and for as long as it is in financial duress to cease paying and/or to defer, reduce, or cancel the payment of interest on any and all loans until one (or it) is in a "not under duress" financial condition; (A "10:59" filing.) (In September 2015 the UN passed a similar law.)

If the Israeli-Palestinian "leaders" had any interest in resolving the conflict they would be meeting weekly with no preconditions. They have no true interest in resolving the conflict because these children of hell are benefiting politically from their posturing in opposition to each other and suck money from others, including over $3 billion a year from the US.

But that's how all politicians from all counties stay in business; by arguing like hell to others over crap while refusing to have a dialog with each other about it, as you know from your own whorehouse government.

Now, let me waste my time addressing the so called "issues:"

    a)   There's nothing special about Jerusalem except that it's a violent hypocritical wh.house in which all of God's prophets and Son were murdered.

Does Dallas argue with L.A. over who should take the full credit for murdering the Kennedys?

Nonetheless I liked Jerusalem and while there I was very excited to return to the hotel in the evenings without having gotten murdered.

If these leaders had any right sense they would each insist that the other side take Jerusalem.

However, it is the Jews that can most correctly claim to have murdered God's prophets so they should keep Jerusalem.

What is the problem with Ramala, for Abbas? Is there something wrong with Ramala? Is Abbas looking to get a promotion into a Jerusalem house, for making a deal?

    b)   The refugee issue is a non-issue also. The whole point of a two State solution is to keep an Independent Democratic Jewish state.

Also, some insist that there is a need to keep all the snakes— who Jesus called vipers— in one place.

There can be a limited number of Palestinian refugees returning to Israel, equal in number to the Jews that the Palestinians let live in West-Bank, Judea.

The American Jews should pay $10 billion for the rest of the refugees to get something somewhere, maybe in Florida!

    c)   The borders are clearly delineated already, in fact there is already a big wall making the old Berlin Wall look small.

The Israelis, as expected, have been cheating and the wall is beyond the '67 borders. So, because that whole area is a dump, they should pay $ 10 for the difference but because that is not going to get the deal accepted, they should pay another $10 billion for the land that they have encroached in.

    d) Security: In my view the whole Middle East from Iran to Turkey, including Israel and from the Arabian Peninsula, Egypt to Morocco should become Demilitarized and Disarmed, if you don't want to worry that all hell broke loose every time you look at the news.

But since that is unlikely to happen soon, at least Palestine should be demilitarized and probably disarmed. Then and only then there is no security problem.

Has Israel denied your right to exist?

For Muslim and Arab governments like of that Saudi Arabia, of Hamas, of Iraq and of Iran to not recognize the right of Israel to exist, irrelevant of how the currently very unjust to the Palestinians, West-Bank/Judea issue is resolved, is contrary to God's Will. It has no Biblical or historical validity, it is very wrong and it lacks common sense. This is one of the first changes that must happen to get to Peace in the Middle East.

The reduction of wrongs isn't a matter of religion or faith but rather it's a matter of our common humanity. It is about not wronging any other human or allowing the unjust suffering of any human as we would not want that to happen to us. The reduction of wrongs is a matter of civility.

The increase in selfless righteousness, in all its expressions, is what religions and faith are about. Hopefully there will be places were the righteous don't have to keep getting punished by societies ruled by wrongful violent-greedy selfish hypocrites who pretend to be civil or "religious" but are "wolves in sheep's clothing." (Matthew 7:15)

One may be in oneness, peace and joy only when one chooses to establish and be in one Independent Demilitarized Disarmed cooperative Holy State, within their nation.

National and Global Security

- Obama should publicly call for and pursue a global Treaty for verifiable freezing at the current levels of the military spending of China, Russia, N. Korea and Iran along with the US. At least he should get an agreement on verifiable caps of military spending increases, particularly of China's increases of 12%/year.
- For the Nuclear Non-Proliferation Treaty to be sustained, the UN Security Council needs to get N. Korea and **Iran** to roll back its nuclear program and then freeze it.
It seems to me that thorough and positive work has been done by the Obama Administration on setting the foundation for the agreement that prevents Iran to have the nuclear weapons that Iran says it doesn't want.
After this deal is finalized or as the next phase of it, I hope that the issues of limiting Iran's ballistic missile range, of recognizing Israel and of moderating Iran's role in the Middle East i.e. in Iraq, Syria, Yemen, Lebanon and to Hamas, will be addressed.

To Presidents Obama, Putin

The only solution that will work sustainably about Iran, Iraq, Syria, ISIL, Palestine, military spending and Ukraine is a solution that both Obama and Putin agree on and commit to.

So Obama should act humbly— as he preached in a recent prayer that was broadcasted— by treating Putin as a "global power" and should offer the solution of, a verifiably by the UN, Independent Demilitarized Disarmed States in Eastern Ukraine, —probably Crimea needs to stay with Russia— and in the ISIL held territory and in Palestine so that the terrified super-paranoid, super-hypocritical Bitch Putin stops fearing and imagining Western aggression.

Even though I suspect that the egos of neither Obama nor Putin want to, the most important action that can be taken regarding all the most serious global threats/issues is for a 2-3 day Obama-Putin summit with the above agenda.

Unfortunately,(as you know) what happens in this world depends on what governments do, which externally affects almost every aspect of your life and has been up to hypocrites like Obama, Merkel, Putin and Xi Jing Ping and not yet up to the Father, the Son and the Holy Spirit.

The world has been ruled by Might not by Right.

Now is the Day that world rule must transfer to Right, the Son.

Now Lord, please let me get back to people that aren't hyper hypocrites.

Because the arguments about Jerusalem also have to do with 3 religions converging there let me clarify a little about each religion.

Now there are some good chances for success with only a few miracles of the Blind Seeing needed!

Islam: Mohammed never went to Jerusalem. He was no more in Jerusalem than I have been to Timbuktu; (in spirit.)

The Koran is not wrong but not because of Mohammed but because someone revolting against the Christianity of the time, a Mystic Christian Gnostic, meaning Knowing God Elli, Ellin, by the Spirit of Angel Gabriel, told him exactly what to say and to just keep repeating it, reciting it.

Mohammed was ignorant, uneducated and illiterate and couldn't figure out anything on his own, let alone complexity.

Mohammed never claimed to have written the Koran, because he didn't. Mohammed recited what he was told and later misinterpreted the Koran.

Instead of sticking to repeating what he was told, as he was told, he moved from Medina to Mecca to chase money, power and vaginas and became a hypocrite having murdered many hundreds and was also a pedophile having married and consummated a marriage with a 9 year old, Aisha. Murdering in the name of Peace and Mercy is hypocrisy, and fn. a 9 year in the name of morality and virtue is also hypocrisy.

Here's a quote from the Koran: Sura 30: "The Greeks *of Mecca*
In the name of the merciful and compassionate God.
[1] The Greeks have suffered a defeat in their neighbour country; [2] but after being overcome they shall overcome in a few years; [3] to God belongs the order before and after; and on that day the believers shall rejoice [4] in the help of God;-God helps whom He will, and He is mighty, merciful.- [5] God's promise!-God breaks not His promise, but most men do not know! [6] They know the outside of this world's life, but of the hereafter they are heedless."

By the way, the Arabic word "Sura" is derived from the Greek word "Sera," that means line, sequence or series. Why are Greeks not allowed in Mecca anymore? The Koran isn't good enough evidence for the Muslims?

Muslims should follow the One God, Elli. If they don't know God they should follow the Messiah Isa, Jesus and be in the Spirit of Mercy and Peace of Islam. If they want to they may even follow the Koran as long as it is interpreted spiritually as it was expressed and not by their materialistic standards. Muslims should not follow Mohammed, the murderous pedophile, hypocrite, who was far from "the ideal man" as many Muslims have wrongly made him— even though Mohammed himself never claimed to be anything more than a common man and one of the many messengers of God— which is making Mohammed into a false idol. For sure they should not follow the horrible Sharia law; and for sure-sure they shouldn't impose any of the punishments of the Sharia law which clearly violate God's Law, the Ten Commandments, and are barbaric and inhumane. Righteousness cannot be imposed; it must be chosen freely. Did Mohammed do the least wrong that he could do and still survive under the circumstances—having at the beginning been chased away by the Arabs and having to escape to Ethiopia— and is he a false or a true prophet?

"The tree is judged by its fruit." So far the results are on the negative side.

Why are Muslims tolerating as if there is some spiritual or intellectual validity to what crazy Muslim groups say such as "Western Education is a sin," like Boko Haram means, and in the name of, murders people. And even if there was full intellectual validity that all Western Education is garbage how can Muslims not be condemning daily Boko Haram's and ISIS' murderous barbaric ways against innocent children? Haram Boko Haram.

Can and will the Muslims change? It is up to each Muslim. Islam if the Koran is interpreted and understood spiritually and thus correctly, with the recognition as the Koran says that Jesus is the Only Messiah, is a righteous religion that was necessary for people who, like Mohammed, are illiterate and therefore could not and cannot possibly fully understand the complexities that exist in the Universe and in life.

The Koran does not dispute at all the validity of the Torah or of the Gospels. It specifically validates both repeatedly. (For example in Ch. 5.68)

The simple truth that caused the need for Islam is that neither the Christians nor the Jews of the time were acting by the love, mercy and justice that the Torah and the Gospels teach.

Ironically Muslims, even in the non-violent versions of Islam, fail most to show by righteous action love, mercy and justice.

"The measuring on that day (of Judgment) will be Just; then those whose measure of good deeds is heavy, will be happy." (Koran 7. 8)

As Christianity and Judaism had to go through a Reformation —and must undergo another reformation—now is the time that Islam needs to undergo a reformation as is being advocated by several wise Muslims including by Ms. Ayaan Hirsi Ali.

The reformation for Islam must include a significant separation between State and religion for those who choose to be ruled by the State and by Justice—as has been done in almost all non-Muslim nations— and the reformation for all is a complete separation, independence, of true believers in God from the "State," under the condition of being Demilitarized and Disarmed and living by Mercy.

Christianity: The first Schism, (division,) of Christianity— after starting to kick out the Gnostics in the 4th century— was in 1054 AC., and was the East-West Schism between the Orthodox and the Catholic Churches which was strictly for political, authority, "control over others" reasons, as have been all other Christian divisions contrary to the Oneness and Unity that Christ taught. That division is neither Catholic, a Greek word meaning "for all," nor Orthodox, that means "correct faith" and should be reconciled fully.
The primary theological dispute is the addition by the Catholics of the words "and from the Son" in the Nicene Creed of Christianity.
Specifically: "And in the Holy Ghost, the Lord and Giver of life, who proceedeth from the Father, —"and from the Son" according to Catholics— who with the Father and the Son together is worshiped and glorified, who spake by the prophets."
The Holy Spirit comes from the Father through the Son, as per John 1.3. The Spirit of Truth comes from Love through Righteousness. Therefore the Holy Spirit comes "**and** from the Son," ("Filioque" is the Latin term) and arguing differently is not correct, is not orthodox; and on the other hand the Pope by claiming infallibility, other than for the Nicene Creed, over all other Christian leaders is a nasty power game that is not good for all at all, it is not catholic, and cannot be tolerated by any True Christian.

(The first Imams claiming infallibility is part of the historical cause for the Pope starting to claim infallibility also. But it was and is wrong.)

I was glad to see Pope Francis, whom so far I like, meet the Patriarch of Constantinople. Hopefully they will both soon agree on the Catholic Nicene creed, which includes the Holy Spirit proceeding "and from the Son"— even though the distinction is a technicality with little substance because the Father and the Son are One— and that the Pope will agree to treat the Patriarchs and the religious leaders of other Christian denominations such as the Protestant Archbishops, the Patriarchs of Russia and of Ethiopia, without claiming infallibility but rather as "first among equals." To the extent that control is the concern of the Pope, asking for veto power over some critical issues is better to falsely claiming infallibility.

Even Apostle Peter, whom the Pope is supposed to represent, didn't have or claim "infallibility" as is repeatedly shown in the New Testament.
Do I need to go through the list of Popes who not only were not infallible but were extremely wrong, including the crappy Pope during the Inquisition and the hypocrite Pope who was "neutral" to the Fascist and Nazi regimes?
These and only these changes will make Catholics truly Catholic, allowing other Christians who want unity but not to be bossed around to enter the Catholic Church and will make the Orthodox truly Orthodox.
As a "Greek," (a name by which the "Western" Italians called the "Easterners" i.e. the ˙Ellins,) I was baptized Orthodox and unless kicked out, I'll remain Orthodox even if in my opinion Orthodoxy is incorrect in insisting in leaving "and the Son," my Lord Jesus, "out of the picture" of the Holy Spirit in the Nicene Creed, even if it is a mute theological technicality and assumed because of the Oneness of the Holy Trinity.
Speaking foreign languages was and is necessary for the transmission of the Gospel to all Barbarians. I love the passion of the preachers who "speak in tongues" but one has to be very clueless of Greek to translate "glossa"— from

which "glossary" is derived— that means language and tongue, depending on how it is used and misinterpreting it to mean physical tongue; and to interpret "speaking in a (spiritual) language that others don't understand" to speaking gibberish "in tongues" and hoping that someone can interpret the junk to make sense of it.

(So long as they don't keep doing it) I will interpret the gibberish that my beloved Pentecostals say in non-offensive 'tongues." They mean: "I have faith except that it's all fn. Greek to me." Fair enough.

The first major theological dispute within Christianity was among Ap. Paul and Ap. Peter as to whether Christians had to convert to Judaism first to become Christians.

As resolved by Ap. James, Ap. Paul was right in claiming that gentiles didn't have to get (physically) circumcised because the Holy Spirit is for everyone.

Ap. Peter was also right in claiming that gentiles had to convert to Judaism spiritually before becoming Christians because Jesus, Righteousness, came not to violate the Ten Commandments but to fulfill the Law, as He said. He did so by teaching people that being loving and righteous towards all others is good, right, worthwhile for you, and is the Spirit of the Law. It's the minds and hearts of Jews, Christians and Muslims that needed "circumcision."

Judaism: The reason the Jews were chosen as Elli's people is because He likes Mary most.

Don't just keep talking about being faithful to God and loving and praying to and honoring Elli but rather live in goodness and do right truly as He Commands by the loving of others Spirit and letter of God's Ten Commandments.

If any disagree with me that is fine. That is why the spiritual world is a quantum/probabilistic world, so that each may have the freedom to be as wrong or right as each wants.

Because the worlds of the spirit i.e. of emotions, of thoughts and of the conscience are also quantum and are invisible directly to the eye, their description is probabilistic, their relationships are correlations, are by synchronous action and the very notice of them changes them.

The solid matter and the physical body causal relationships need have no exceptions (such as chairs have one or more legs). Unlike the solid physical because the existence and description of the spiritual, intellectual and emotional is by probabilistic waves, their relationships are not causal but are correlations and have significant exceptions.

So what I said about the Europeans, women, Jews, blacks, Palestinians, Muslims and everyone else that I hope to have offended to Wake them up, isn't really about them because for the most part we are all the same and on the whole mostly in goodness but it is about the very small, smaller than 1% of 1%, at one end of the curve, who lead and exploit our, otherwise beautiful and very useful small differences.

There are many, the vast majority of all humans who are kind and trying to do right despite the price they pay and they have been blessed.

Also, to rebalance possible misconceptions by my railing against lawyers and the US, I should mention that in my judgment despite the very many serious flaws mentioned that need correction including in the US legal system, it is still much better than the rest and was a key reason that I immigrated to the US and am still proud to be called an American. The US is still the last best chance for humanity.

There are kind, trying to do right, people even higher up in the hierarchies, for example, even at the top of the entertainment hierarchy and even among Jews, there is (there was?) Jon Stewart who is... serious, and Lewis Black whose excellent expression of anger I tried to imitate in the next Chapter.

If you think that I am wrong or not true somewhere, prove that you are right. Prove it by results. The results of being right are peace and God's Kingdom on earth. Show me.

I am Imprisoned, Naked; Homeless; Hungry.

Please do not do to me as your forefathers have done to poor tragic Homer, Sophocles, Socrates, Plato, Moses, the prophets, Buddha, the four (Ellin) Mystic Aryan Vedas who founded Hinduism, Lao Tzu, who left that society altogether after writing his book, the Apostles, the saints, and Christ Jesus, which is to find what they say admirable, right, honorable and honor them in words but only after by your apathy they die so that they get no satisfaction while in this world and then keep remaining clueless as to what they said, not understanding what they meant, why they said it and what it is that they hope that you will do and doing it. Stop doing wrong to yourself to improve your well-being and the state of this species.

I expect that this book will be broadly liked after I am long gone from this world. If you do to me what you did to Christ Jesus, the prophets and the saints I will f. you even if I have to wait until after this life.

It felt like the great Ones pleaded vulgar me— the Greek word is Parakalo— to explicitly explain their teachings.

Their beautiful, non-offensive and proper teaching can be understood by the vulgar people (like me) as follows:
This nice looking materialistic world is spiritually a mean, immoral, violent, incompetent, ignorant deceitful, unjust hypocritical Whorehouse that you must get out of fast, spiritually first and then physically into a non-violent, Disarmed Demilitarized Democratic Independent Holy Land, State or die tragically.

Neil deGrasse Tyson is currently a famous astrophysicist, for popularizing science and for having been critical in "demoting" Pluto. Pluto means wealth and is the same component word in plutocracy (by which you are ruled whether you know it or not.) Unlike all the other planets that were named in English by the Latin names of Greek gods, Pluto somehow got in there also. It is a good thing that fn.

Pluto was demoted. Pluto, wealth, money, Babylon and their pretend, false god Mammon are narcissists with undeserved high opinion of self!

There is a beautiful image of the earth as seen from a long distance that N. Tyson shows during his "shtick" and calls "the light blue dot." (It was originally used by Carl Sagan.) That very long distance prospective in terms of size and of thinking in terms of thousands of years, as astrophysicists need to, is a correct prospective.

From that correct prospective of being on a blue dot for a short "day," one is forced into the conclusion that all the current disputes, acrimony, bickering, fighting, angst, stress, violence and wars are all senseless, as Mr. Tyson points out in nicer language, starting with calling for a time-out.

Not taking care of each other and the "blue dot" and not enjoying our brief time on this blue dot, makes no sense.

Don't waste your extremely little free time on this blue dot in doing anything else other than establishing the option of an Independent Demilitarized Disarmed Democratic State in your nation so that you, others with similar prospective, and children may start fully enjoying in Peace, Love and Liberty your very short, fleeting time on this blue dot.

The current hypocritical, violent, whorish wrongful, selfish, stressful, self-damaging, painful, deadly, morally depraved condition of humanity can have only one defense and it was given by Jesus just before He was murdered:

"Father, forgive them; for they know not what they do." (Luke 23:34) I believe that indeed people must not know what they do because if they do know and the results are what they have been then those who either claim or do know what they do are surely guilty.

Well, is it not time for the people to get some truly correct education and so to truly figure out the impact of what they do?

13. Education Olympics

The current education system, by testing that requires one to pick the correct 1 out of 5 or 4 answers fast, rewards those who identify fast the correct data that is surrounded by 75%-80% of intellectual crap. Yet, that capability is useless here because there is none of that.

I dislike profanity so I tried to write this book without using any profane words but found it impossible to do because these glorious humans have said everything nicely already. The only thing that was left to do is to make their Teaching explicit with profane language so that even profane people, like me, can clearly understand and relate to what their own chosen, trusted Teachers taught.

Alma-Mater is a Latin word, stolen from the Greeks and then distorted, intended to mean, spiritual Mother.

Dear Harvard Alumni: "Veritas," Truth is what we stand for.

It was reported that Leonardo Dicaprio came out of a bar with 20 women; and so he offended Dionysus!

To avoid WWIII and Armageddon, you are each urgently needed to join us Greeks in Spirit, Intellect and courage by our Greek fraternities and sororities to defeat once and for all the ignorance that is causing pretty single women to suck the wrong penises!

As you know our major enemy, the ones that call miseducation education is fn. Yale.

It is those alumni from Yale, aligned with the barbarian Romans and Germans that keep stealing our jobs, well-being and women. We must beat Yale and fn. Princeton once and for all, so that there is no question in any woman's mind that only people from Harvard, Columbia Greeks and most importantly you rightly deserve and should get your genitals licked and no one else!

Fn. Yale and Princeton are misguiding the people because they have deviated from the advice of their first and best Educators.

Jonathan Dickinson, the first President of Princeton said: "Cursed be all learning that is contrary to the Cross of Christ."

In Yale, Benjamin Silliman wrote: "It would delight your heart to see how the trophies of the cross are multiplied in this institution."

To avoid WWIII and Armageddon I declare Education Olympics to be held on 10/1/ 2015 or whenever you get to it, in the Baker Library of Harvard, in Boston Mass.

The rules: There is only one question: What are the relationships, connections, correlations and causation within and between the following 12 disciplines?

The 12 disciplines, that are explained in the Education Exhibits and their correlations that everyone should know enough about, before graduating are: Physics, Thermodynamics, Chemistry, Biology, Psychology, Business, Business strategy, Investing, Economics, Politics, Philosophy, and Religions.

This knowledge is essential because without it one cannot know their body and its functions that are necessary for physical survival or know what is necessary for their emotional, economic, political, intellectual and spiritual survival nor understand and integrate the critical relationships amongst these disciplines. This understanding is necessary to fix the problems in people's lives which invariably are both physical and spiritual.

Each and all of these fields are critical determinants of all things, of each life and of all life.

These disciplines are about and are as interrelated as your body, emotions, mind and soul so that those "experts" who understand and offer advice about any one of these disciplines without understanding the impact on or from the other disciplines are wrong because they are bound to be wrong about the whole.

One member, either student or faculty or alumnus from every University that chooses to participate from anywhere in the world will be chosen by that University and sent to compete.

The judges will be from some of the Universities that do not participate as competitors led by MIT, Sorbonne, Oxford University and the University of New Delhi.

Participant Universities will choose to compete in one of two teams.

The other team is to be led by fn. Yale, Princeton, Stanford, the Barbarian Bavarian Munich University, the University of Moscow, the Baloney University of Bologna, all the Universities of China and of the Muslim nations, Leo, Hollywood, all the political leaders of this world and their governments who may argue that some of the teachings of the Sciences and Humanities are inconsistent with the Cross of Christ.

Consistent with the wish of President Dickinson, this team shall be called ignorant mother fuckers; (operating by the oedipal complex.)

Our team will be led by you dear brilliant friends and from Harvard, Columbia and Greeks proving that the teaching of each and all these disciplines is consistent with the Cross of Christ; as is shown in the Education Exhibits and we shall be called truly civilized and beautiful.

Oh! I almost forgot all the Rhodes Scholars go in the "other" team as well because it is a well-known fact, according to my dearest dad Andonis, who married a wonderful lady from Rhodes, my mother Flora, whose parents my dad didn't like but because he didn't want to blame them directly (and to excuse them) he kept informing me with a pitiful look that all the people from Rhodes are fd. up! So, Andoni, for you: fk. the Rhodes Scholars also; they go in the other team!

Our numerical odds, by my calculations are very respectable 300:1 odds against us so that no one may accuse us of beating up the weak. The odds against us are respectable enough to get every hero out of bed and even some lazy Greeks out of bed. And those odds include my other alumni brothers and sisters of the National Polytechnic of Athens and of the Hellenic School of Addis Abeba.

Of course, us being civilized and doing what we can to avoid being offensive, we will have to lie as per the instructions of Homer to flatter the beasts of the other team and tell them that they look civilized and we will even publicly call them the "looking pretty" team even though we can clearly see, (like Shakespeare's Merlin's owl, Archimedes, that sees through the darkness,) that spiritually they look like horribly ugly mean monstrous wolves, beasts, zombies, dragons and vampires.

Those who try hard to look pretty on the outside are often broken on the inside.

After the preliminaries are finished there will be 1 to12 members in each of the two teams. In the team competition each member of each team will make a presentation that supports the team's argument for their academic discipline.

The knowledgeable beautiful people of our team will make the argument that the findings of their discipline are consistent with the Cross of Christ, as shown or similar to the Exhibits.

The ignorant pretty looking mother fuckers of the opposing team will argue that either the findings from their discipline are inconsistent with the Cross or are inconclusive.

In the individual dodecathlon competition each participant will make a presentation, consistent with their team's argument about their knowledge and correct understanding of each and all of these 12 disciplines.

Then they will be questioned by other competitors, judges and will provide answers. A short debate will follow. After the debate is finished, the judges will make their judgment.

When the Olympics are finished and the judges declare unequivocally that we win, then many women of this world will be Resurrected and realize that the men that are good and right for them are not those who look good and right by their wealth-power and fame who are usually hypocrites like "whitewashed tombs that look clean and beautiful on the outside but are full of filth and ugliness on the inside"

(Matthew 23:27) but those that are truly good in their spirit and are right or correct in their reason and in their actions.

So fear not dear friends the huge odds against you because every day is a beautiful day to die away from this world and fighting for Truth and Liberty is a beautiful way to die.

Those who believe in power, particularly the Germans should know what F. Nietzsche, who was the intellectual leader of those who believe in Power, wrote:

"Hence confronted by the Greeks, people have been ashamed and afraid, unless an individual values the truth above everything else and dares propose this truth: the notion that the Greeks, as the charioteers of our culture and of every other one, hold the reins, but that almost always the wagon and horses are inferior material and do not match the glory of their drivers, who then consider it amusing to whip such a team into the abyss, over which they themselves jump with the leap of Achilles." (The Birth of Tragedy: Chapter 15).

Consider a dying patient and the doctors refusing to help her because of "moral hazard." The morally right is called moral hazard by this world's whorish economic leaders of governments. I watched a press conference by a son of shit W. Schauble on April 18th in which he warned of the "moral hazard" of doing what is right for Greece, (while Germany has been the external cause of Greece's economic collapse.) I admit to getting enraged whenever cowardly racist barbaric evil whore scum Germs like him dare talk about Democracy, ethics, morality, Greece or Ellas.

It is the hugely immoral barbaric whores like those leading the Germs of Germany (where whoring is legal) who have done nothing moral, have been most destructive to others for their own self-interest and are incapable of being moral, like the bound to the Abyss scheisse W. Schauble, who not only don't do any good for anyone else but call any such right action, (any "doing good") a "moral hazard."

There is IMF precedent for debt write-offs predicated on some reform and so it is not a "moral hazard."

The most significant reforms have already been made in Greece and include having already a primary budget surplus, and the political reform by the Greek people in having scrapped the 3 political Parties that put Greece in bankruptcy.

The main reform still necessary for Greece's economy is an external package of foreign direct investment along with an internal package of incentives for industry (for areas outside Athens) and for high tech foreign direct investments and for exports. That package should include multi-year tax breaks and suspension of the restrictive labor laws (and hostile Unions) in exchange for some minimal local ownership, like 10% ownership by executives (which is how most executives make most of their money) and 10% by non-executive employees.

This should also be part of privatizing some public companies.

WEALTH DISTRIBUTION:

Companies and the working people can start doing well (without political "connections") in Greece (and anywhere else) only if there is local ownership not just by the local investors by also by the executives (10% or so) and 10% by non-executive employees.

Competent executives know that their most valuable asset is their employees.

In the last company I run, I was able to negotiate from my Venture Capitalists only about 7% stock option plan for the non-executive employees but it was enough to produce over 230% per year returns to stockholders, that included all employees also.

This is the only way to reform the otherwise not changing labor laws and the mentality of seeking a permanent job in Government in exchange for votes for the political Parties.

Also, some measured compromise to increase government income from shipping (and/or ship owners) must be reached.

The 40% write-off of Greek Debt could be predicated on these last reforms that are needed.

The single most important law in developing nations to both grow and to protect their labor, resources and culture is to require any public company to have at least a 10% of the subsidiary within that nation 5 year stock option plan for the local non-executive employees and within a year, a seat in its Board.

Governments should treat the profits of non-executive employees from the sale of stock options as capital gains, not income (expense), to incent corporations.

This and only this law will allow growth without exploitation of a nation's labor and resources.

The US should make this change to the IRS rulings and offer this incentive and possibly some of the other incentives that corporations with a 50% ESOP plan have to any corporation with a 5% (possibly even 3%) stock option plan (beyond the 401-k) for non-executive employees in the US.

After that requiring through the SEC that these (American non-executive employee) shareholders should also have a Board seat is a short and reasonable step.

Only this law helps align incentives, over the long term makes the job of executives a lot easier, is beneficial even to small independent shareholders and will generate both high growth and high wealth distribution.

This is the best anti-poverty and pro-middle class propelled growth program within these systems at this time.

The second best alternative to reduce income inequality is to expand the earned income tax credit which is better than increasing minimum wage which has some negative effects on businesses and on economic growth.

When the judges declare you the unequivocal winners of the Education Olympics then the Mafiosi Italians will get sense and like George Clooney marry and become respectable because Mrs. Clooney immediately after her marriage called for the return of all looted Greek Sculptures from the Parthenon in the British Museum back to Greece; proving that Mrs. Clooney is a good Virgin in Spirit, Lady; and finally George Clooney stopped pissing us off!

The winners, having explained everything and Life in the Universe, logically, with evidence and consistently with the

Cross of Christ, will be honored for truly knowing everything and all, (without pretending so while being clueless and not knowing what they do, like the hypocrites,) will win the competition price, which I advocate should be $310 billion and 21 pretty looking women for the night (fn. Leo!...) for each of you.

The cheerleaders for athletes (who can't even qualify for any Olympics,) Leo and fn. Hollywood should find and deliver and the 21 volunteer lady trophies for each of you, along with their apologies for their past transgressions.

And the Sirens like Katy Perry who sing about "manage a trois" every Friday and haven't invited you— offending Dionysus— should invite you next time reader friends if you aren't hypocrites about Truth and do something now for Education Olympics to communicate concisely and as simply as possible the whole truth to all.

All those who pretend to know good and right or who say that they are right or that they do right will be then proven wrong, ignorant hypocrites and will have to pay the price.

Lawyers aren't necessarily bad in this context. In the context of this world a central difference between good politicians, thieves, prostitutes and lawyers and bad ones is whether they're in your team or in the opposite team. The best lawyers if they are in the opposite team become the worst lawyers.

However to be fair to the ignorant pretty looking team, lawyers will not participate in the arguments because the arguments are scientific and philosophical and are not legal and because if we included alumni from Law schools, given that our team would have 6 out the 9 US Supreme Court

Justices, while Yale has only 3— mostly for reasons of affirmative action— we might as well declare victory upfront and stay in bed. In this "Just and fair" system of law that we live in, discrimination is not allowed anywhere else, except that all other Law Schools in the country are obviously irrelevant, as far as the Supreme Court is concerned.

I know that I am not supposed to say anything against my team but, in Truth, I belong only in God's team. I sure hope that in classes other than the one I took in Harvard Law School they teach the proportionality and harmony required in law, given that this society is ruled by the law.

Being able to understand and to make the best arguments for both sides is inadequate unless one understands, why and how to then argue for the proportional, geometric mean, between them, as per the preponderance of the evidence, that allows for harmonious resolution.

As to the Harmony between the four basic conflicting arguments of every case, Beethoven captured it in his 5th Symphony. The Big Bang was not a single Bang but, as we know, it was a sequential release of the 4 forces; so that it was Bam ba ra baam. That beat and a corresponding geometry permeate the Universe. It has been detected and measured, by COBE, as a quadro-polar Universal background radiation.

If you, Harvard Law and Supreme Court lawyers do not understand, teach and practice the correct proportionality and harmony, you produce cacophony and imbalances not just in the law but also in the whole society that you operate by the law. The US Constitution sought the correct checks and balances and is beautifully balanced. Are your opinions, individually and as a whole correct and well balanced?

Despite what Obama says: that ISIL is not Islamic, which is a lie with diplomatic and political validity, **Islamist** extremist are the only ones for whom there is any probable cause to search their "effects."

The "solution" of "search and/or seizure," through filters, of all the digital "effects" such as private e-mails of all US citizens is unreasonable because there's no probable cause that all the people are the enemy of the State. This has been a violation of the 4th Amendment as the District Court ruled.

Instead of persecuting Ed Snowden for revealing that there is a serious violation by Bush and by Obama (who campaigned for reform of the Patriot Act and then secretly expanded it) of the 4th Amendment, Snowden should be given asylum and honored.

Whether one can access the content (the data) through the metadata is critical but unlikely to be revealed by the NSA. Both Republicans and Democrats agreed that the expiration of the Patriot Act was another "unnecessary manufactured crisis" and then the Senate passed the "USA Freedom Act" which transfers the bulk collection from the government to private companies.

Both sides agreed that this new law will probably not work because some private companies already said that they will not keep the metadata unless required by law and the Freedom Act doesn't! (If this wasn't real it would be funny!)

In the next re-authorization of the "USA Freedom Act" it would be useful to have a minimum and a maximum time of how long private companies may keep people's metadata because if not the Act isn't usable and it can be a source of serious damage to individuals and the nation.

Also, the opinion that corporations (which are no more than paper) are people and that money is speech (Citizens United) is unreasonable, is inhumane, has the effect of a serious social imbalance and is making this a plutocratic—the rich rule— Babylonian nation.

Dialects evolved and eventually languages separated by people who didn't want others to understand them. So is legalese; is your law about whether a particular case is more akin to werenottelling v. gibberish or to letsbenotunderstandable v. fkthepeople?

As to the in legalese opinions that no one from the public reads: no judge that isn't a hypocrite makes any judgment without stating upfront the criteria used for their judgment so that if others use the same criteria, they would end up with the same judgment.

Right Judgment requires more than whether a case is like fuckyou v. blome; or rather to blowme v. f.you!

Would you say that an Independent Demilitarized Disarmed Democratic State within the US is Constitutional under the not "impeding the free exercise of religion" of the 1st Amendment or not?

Even if an Independent Demilitarized Disarmed Democratic State was established within the US Constitution would it be able to withstand the challenge of some within it, arming them-selves on the basis of the 2nd Amendment? Advice on this issue would be helpful.

Since you don't deal with hypotheticals, let me give you a real case. My conscience does not allow me to accept any government that in the name of protecting me commits even a single murder of an innocent civilian either intentionally or as an unintended casualty.

There are too many murders, casualties and severe abuse of innocent people, here and even more abroad, by these governments in the name of my protection that my conscience and religion does not allow me to accept.

I therefore declare for reasons of conscience and religion that have been explicitly expressed in this book, my privately owned Land in California, an Independent Demilitarized Disarmed State, independent of all local, regional, national and global authorities. I intend to relocate and expand the Independent New State and seek its recognition as a sovereign and Independent Demilitarized Disarmed State.

Is this Independent, maybe even from you, Demilitarized Disarmed, Democratic State Constitutional or is it not?

Lawyer Competition: If lawyers want to compete on a side competition, the question they need to best answer is: How can an Independent Demilitarized Disarmed Democratic State be made Constitutional; and if it can't what is the easiest process for a State or part of a State to peaceably secede?

Petitions to secede have been filed in all 50 States and most polls show that about 22% of Americans want to secede.

No State has a majority to secede but if all of you who want to secede, for whatever reason, concentrate in asking for one small Independent Demilitarized Disarmed Democratic State you would have more than enough votes to peaceably and successfully secede from a State and then from the Federal Government. If succeeding in seceding is important enough then moving shouldn't be too much of a huff.

Even those who interpret most "Conservatively" the Second Amendment should find this compromise acceptable and support one small part of one State for their "enemies," thus leaving all the rest of the country assured about their gun rights.

Even though the lawyers will not compete in the Education Olympics but only as a side show to find out if they are indeed impeding the entrance of others into God's Kingdom— as Jesus said they have been doing— and those who do impede it are condemned to hell, any and all unresolved disputes will be resolved by the US Supreme Court because that is the highest Court, so it must be treated as right even when it is wrong.

The financial proceeds from the EDUCATION OLYMPICS will be managed by the organizers and should be distributed as soon as possible follows:

- $20 million should be allocated for the organizers, to organize, fund, promote and for participants of the Education Olympics.

-$20 billion to the Palestinians; $ 10 billion to the Ukrainians, if and when they agree to have sustainable peace; and $ 9.98 Billion to fight the war on ISIL in Syria and/or Iraq and to establish there one "safe zone" Independent (mostly Sunni) Disarmed Demilitarized State, for the refugees and others, hopefully led by women.

This is also essential as a refuge for hundreds of millions of Muslim women that are under really brutal slavery today and as a refuge for the Christians in Muslim nations that have experienced and are experiencing genocides maybe as great as those of the Jews by Hitler.

-$100 billion for establishing, constructing and operating a Small Independent Demilitarized Disarmed Democratic State within the US.

-$170 billion to pay the debt down of the fn. Greek government. (Actually it's a paper write-off by the ESM.)

The Debt/GDP ratio of Greece went from having been stable for decades at 100% in 2001, which is the barely sustainable limit of any economy, when it entered the Eurozone to 175% in 2013, while the Greek Governments were being told what to do and managed by the Germans.

And this is supposedly after the "charity" of the Germans to force all Greek private owners of debt to take a 50% or so cut on their investment and the "charity" of refinancing the debt at rates much higher than Germany gets just because it is in the Eurozone.

So the **defeated German** germs **that have gone bankrupt twice and begged for charity** for the children of the mass murderers and were given that real charity twice, including by Greece, have now enslaved, have caused much suffering and have unethically and immorally hugely indebted Greek children for generations to come.

Greece owes nothing to Europe or to the world; rather Europe and the world owe to Greece. If you are a German and do not revolt against your government, as some have, to write-off some of what it stole, which is everything that Germany has, you should be thrown into the Abyss.

Or else the Greek government should go to bankruptcy proceedings through the UN and come out with 0 debt.

The Germans (may physically fk. all top Greek officials if they like; I don't care, but they) cannot, should not and will not burden Ellin children with an impossible burden without burning in hell forever for it.

This debt write-down is not any real money transfer, it is just a balance sheet "one time action" that will have no negative effect on any economy, will not have to be paid by any taxpayer (as the German government lies about,) will not increase the debt of any nation and will actually improve the economy of Europe having put that "one-time write-off" behind.

The economic war and the next Debt-currency Crisis

The total trade deficit of the US in 2014 was $505 billion. The trade deficit of the US with China in 2014 was $340 billion and with Germany $73 billion, so that China and Germany make up over 80% of the US trade deficit.

By far the biggest heist, the biggest legalized thieving of all time is happening right now by a clan of executives in Germany and by the members of the Communist Party of China who are stealing hundreds of billions of dollars a year from the US, the UK and from Southern Europe.

One finds similar statistics of big systemic trade deficits by the UK and by Southern Europe to China and to Germany.

If those trade deficits were getting rebalanced as they are supposed to by the market mechanism of free exchange rates, these deficits would not be a problem; but they are not getting rebalanced because both Germany and China have established mechanisms to keep their currency devalued.

Both China and Germany have a clear mercantilist strategy i.e. intending to bankrupt other nations.

It is done through multiple mechanisms including currency manipulation, outright intellectual property theft by China and outright corrupt laws in Germany that allow German multi-nationals to expense bribes in foreign nations, protect those that received the bribes within Germany, and then accuse foreign nations of being corrupt... by envy and corruption that the German companies incite by offering bribes.

Lao Tzu called the "incitement to envy" the worst of sins — in my judgment killing is by far the worst of sins but to someone seeking divinity it is understandable to view the incitement of envy— as "dreadful as envy/coveting."

My first job out of HBS was to manage for GE Medical Systems where I saw personally how many foreign markets where "locked up" by briberies by German companies like Siemens. (This bribery German law may have been changed since.) Because these trade deficits are not getting rebalanced by free foreign exchange and are in fact increasing and have been increasing for over a decade they are bound to cause debt crises and recessions in the trade deficit nations.

It is these trade imbalances that were the external cause of the debt crisis of 2008 and I am shocked that it is not widely understood yet.

The international bankruptcy law for nations was tabled by Argentina to the UN and was supported by over 120 nations but failed because it was wrongly opposed by the US and the

UK who would be wise to change their position on it because they are the most likely to need it, next.

Obama and Cameron must insist that Germany ends its abuse of the common, fixed, currency with Southern Europe and **ends the European debt crisis** by using the EFSF and other instruments for the public sector to write off now at least 40% of the debt of the Greek (as they required the private sector) and 10% of the Portuguese and Spanish government debts to bring them back within the original debt/GDP requirements of the (neo-Roman-German Empire of the) fn. Eurozone. The troika of the fn. ECB, of the Ignorant Mother Fuckers of the IMF, and of the fn. EC should be kicked permanently out of Greece.

Unlike the open economic war through sanctions that the US is waging against Russia, Iran and N. Korea for political reasons, the economic war by the big corporations of Germany and by the Communist Party billionaire officials who own the Chinese companies is hypocritical, is done while pretending to be friends or "frenemies," in cahoots big US companies.

The Conventional economic Theory says that the trade surplus nations can and should increase domestic spending and domestic consumption. That is what the free market mechanism would do if these currencies (the Deutschmark and the Yuan) were allowed to appreciate and this would help the lower and middle class of Germany and of China.

But that Theory of "increased domestic spending" by China and by Germany will be done very slowly and as a wasting time mechanism, because in truth the governments/big businesses of China, Germany have a clear intent and a coherent strategy to bankrupt other nations and to oppress their own lower and middle class, despite their lying words.

There must be import taxes, as even Adam Smith and J. Keynes argued, on those who pursue mercantilist policies of sustained big trade surpluses.

There is no way the US can keep "leaking" with two huge "net" "leaks" of over a billion dollars a day and at some point not sink, through another debt caused recession, as is also the case for the UK, Russia and Southern Europe.

The import taxes by these governments to both Chinese and German goods must be linked to the trade deficits and go up and down as trade deficits from Germany, China increase/decrease.

So when the next over-consumption, debt caused recession happens even though the external perpetrators are the Chinese and German top brass who are hopelessly hypocritical evil thieves, the people who were supposed to defend you and keep failing to do so, Obama, Cameron, Lagarde, Holland and the current top political leaders of both political Parties are the ones who must be internally blamed.

The US is currently funding through deficits China, which is by far the biggest political adversary against Democracy and the biggest economic (and cheating) competitor and biggest potential military adversary of the West. The US is relying on China to build more empty cities to move its poor rural population into those empty cities for global growth.

-Instead of funding China to build more empty cities for growth (as m.fkn. bankers like H. Paulson want for big commissions) why not build one and only one new beautiful technologically advanced rural, suburb like, environmentally, architecturally, socially, economically, politically well designed and well organized Independent Demilitarized Disarmed Democratic State, here, to move the poor from your inner cities, the 70% of your imprisoned who are non-violent offenders, the refugees and the blessed to help provide economic growth for them and for the global economy?

Why not build a New Independent Demilitarized Disarmed Democratic Holy State so that you not only set the right course for humanity for now and the short and longer term future but also to set the course of humanity right for the millennium and be honored forever?

Dear friends, establish these Olympics of Education to avoid humanity perishing by its ignorance.

The recommended seed investment to lead this whole project, that will also better balance the portfolio of the endowment funds of Yale and of Princeton, is only $310 million.

If you don't do the Education Olympics, as I advocate them, I will be forced to declare Armageddon and go to war into WWIII. I teach all other disciplines but I don't teach war because I do war. Other than not learning anything about war from me, if you hoped to learn about art from this book you are also out of luck! The concept of Art escapes me!

I can't analyze; synthesize or induce into a generalization; deduce (apply to a specific case) anything from it, other than bam baa ra baam; Let There Be Light.

I like art very much, surround myself with it, and have studied it but unfortunately I can't get it. My brain cannot process how to judge art objectively. Painters mix beautifully the 3 primary colors that make up the top 3 axes of light of each Cross but how they do it beats me!

So if anyone says that "this is great art" or "this is the best art" no matter what I think of it, even if dislike it, I have to politely act like a dog or slave and go along with them! The only advantage I can think of me being so utterly unartistic is that it may help increase your confidence that you are a brilliant artist relative to me.

So, if you are an artist and need resume material you may say that the author whom the people are eagerly waiting to make extremely famous soon after the bastard dies has given you a recommendation letter, this book, which says clearly that relative to him, you   are an incredibly brilliant and beautiful artist.

The art I find most interesting, beautiful and instructive is children's art.

Talking about what this book cannot offer, I heard that there is a thing called Emotional Intelligence; God refused to give me any of that! God specifically must have said, "don't be greedy; no emotional intelligence for you," so I have no comments about whatever that thing is!

Harvard Business School, HBS, uses case studies and Socratic dialogue to teach. The cases are about a real decision point, usually with some difficult to choose from options, in someone's business life.

HBS insists that there is no single right answer for any of its cases. Why? Because there is uncertainty in life and people do not know for sure what will happen so any choice could be proven wrong by time.

Also because, there is no way to make a right choice between options that include undesirable trade-offs and therefore are in part wrong for someone, as all real life cases. Any choice among wrongs is by definition wrong, even if it is least wrong.

What can be done about cases and about business, economic and political decisions is to be least likely to be very wrong and to be least wrong and coherent and thus correct; with an answer in which the objectives-strategy-structure and tactics are straightly aligned.

There is more than one correct coherent answer to each case. That is because there is a range that depends on the confused, confusing, too greedy narcissist (what's the common name for them?); the over-controlling attention seeking territorial hysteric (doggie); and the short sighted emotionally unstable (empty-angry- sucker) of each case.

Thanks to Homer I didn't have to read or study in any study group any of the cases to know what the usual problems and solutions were, as was explained. (So, I used my time wrongly, "practicing" in the bars of Boston.)

Confucius said: "A man who committed a mistake and does not correct it, is committing another mistake."

Recognizing and admitting— as most political leaders rarely do —that one is inevitably in part wrong in every political, economic and business decision, is critical to correcting the errors they made fast and in doing so to being least wrong and thus correct.

To foundationally improve the governance and functioning of public corporations and as a result of modern societies there is a need for the SEC to require that public corporations add a, by non-executive employees nominated and elected Board Member. Preferably the non-executive employee Board member of only public corporations should lead the Ethics Committee and be in the Compensation Committee.

High schools should change their curriculum and teach less of the details with less homework and more of the big picture of the major Academic Disciplines as shown in the Education Exhibits. Schools should also include one day a month of case studies taught through Socratic dialog.

If you don't implement the Education Olympics I will be forced to prove— well, I proved it already— that all the scribes and particularly the leaders that rule this world and more particularly the heads of Satan that I named above and including Obama, —the black horseman of the Apocalypse after whom the Chinese yellow horseman and death comes— have been very to extremely wrong, deceitful, violent, ignorant, incompetent, immoral, sinful, blind, hypocritical mother fuckers who are most self-interested and falsely pretend that they know and are right while billions of people are suffering from their very wrong actions and inaction.

To the Presidents, Chancellors, Prime Ministers and Kings of all nations:

Bitches, (according to Homer I should say) pretty bitches, there is an urgent need, it is right and it is achievable to raise $310 billion to be distributed as above; and to recognize at least one small Independent, Demilitarized, Disarmed Democratic Holy State, governed mostly by women, anywhere in the world, whose Independence and security is guaranteed by the UN Security Council; which until now has been ruling the world wrongly.

The amount above is about one third of the amount that was used to bail out your fn. Banker thieve friends, to help you and them out for having screwed everyone selfishly, badly, blindly in broad daylight. Import taxes on Chinese products even after considering that they drop to 15% after negotiations and for limited products and the 15 cent/gl. gasoline tax, can raise these funds, from the worst offenders, in about 3 years.

Support for debt relief for Greece and the demand on Germany to stop its destructive to Europe mercantilist policies is very broad throughout the world with articles in many newspapers including in the Washington Post by Harold Meyerson. Even Obama says that he supports stopping the austerity programs in Greece however he hasn't done anything about it yet.

Keeping the Greece debt crisis prolonged and ongoing causes prolonged devaluation of the Euro and thus increases exports and growth of Germany and Holland at the expense of not just Southern Europe but also reducing US growth. Greece is the cradle of Western civilization. It is expected that Western mother fuckers will be trying to fk. Greece. That is why the blood sucking mother fucking great whore Merkel continues prolonging the fn. of Greece.

THE SOLUTION TO THE DEBT CRISIS IN GREECE is simple.

Babylon the Great, Merkel required that the previous "private" creditors take an over 50% haircut and imposed a troika of her European whores (to avoid direct responsibility by Germany) to rule Greece for 7 years so far, for that loan.

Now is the time for the Great Whore who sits on (the Dragon based Draghi,) Eurozone Multi-headed Beast, Angela Merkel, to take a 40% haircut within the Eurozone or a 50% debt write-down with a Greek Exodus from the Euro, absorbed through the ESM, that was set up to avoid any direct damage to Germany, which is kinder than how she treated the private (previous) creditors.

If President Obama doesn't do so, now it will one more huge foreign policy blunder, because Putin will and Europe will be torn apart further, again.

I don't care how the funds are raised so long as it is done ethically. If it's to be done fairly and considering the current wrongly generated imbalances— in whatever mix of public and private financing— it should be about $100 billion by Germany mostly, northern European States and Japan; $90 billion by the Gulf States and Turkey; $70 billion by China; and $ 50 billion by the US who should be first to commit and to lead, (as shown above) to actually solve the serious problems that all humans face and the lack of right understanding of all the Truth from which all humans still suffer.

(My apologies to your dog(s) for calling these "leaders" Bitches!)

False philosophers

The purpose of philosophy is to discover and communicate the logic by which one and all may live fulfilled sustainably joyous lives; ("Eudemonia" is the old Greek term); and to *prove it*. Since Plato and Aristotle —with the exception of J. Locke, H. Thoreau and Emerson— all philosophers have been wrong, false philosophers. Some as Marx, Nietzsche, Hobbes, Machiavelli and the Nihilists, who caused serious harm and deaths unfortunately called their junk philosophy. Others, like J. Mill (utilitarian,) Voltaire, I. Kant, Hegel, F. Engels and A. Camus imagined not proven but at least less wrong stuff.

The current Academia in Philosophy is pre-occupied with Linguistics (the study of language) and with Etymology—which cannot be done well without understanding and speaking Latin and Ellenic-Greek—to add footnotes to Plato. Modern "philosophers" found no right logic yet; they haven't proven anything and so they've given philosophy a bad name.

Therefore, it is better to categorize this work as Unified Physics and/or Unified Science and/or Unified Theory of Life and/or Practical Psychology/Psychiatry and/or True Self-help and/or Revolutionary Sociology and/or Ethical Economics and/or Correct Corporate Governance and Coherent Business Strategies and/or New Correct Politics and/or current violations of the Constitution and/or Right Religious Theology and/or True Spirituality and/or The Correct Education, as it is all these.

It's even better to give this script the label "one deep rap" rather than Philosophical for anyone to want to open it up! Etymologically, the "logy" part of Psycho-logy, Socio-logy etc. is derived from the word logic, and it means the search for logic in a particular field.

These fields were created recently because philosophy had failed to explain coherently the logic describing the reality of these aspects of people's misery.

To prove any philosophy, the true philosopher must understand and explain logically the logic of the laws of Nature thus making science necessary. Unlike all other philosophies, this philosophy, by the Cross, is proven by all evidence.

This self-evidently right philosophy of Right is proven by all the currently confirmed *scientific* evidence as evidenced by the Exhibits; and it is proven by *Biblical* evidence, as validated by the above quotes from the founding documents of all seven righteous religions; and it is validated by *experiential* evidence i.e. by personal experience, by confirmed evidence from others and by historical evidence.

Shouldn't the Professors at least know the Cross of the Discipline they profess?

Philosophers believe that only by changing one's and other people's thinking, one may make the improvements necessary to get to sustained happiness and Joy.

Why do people think wrongly and what are the correct changes in their thinking that are needed to get to sustained happiness and Joy in Peace?

Yesterday, a lady (Mystic) reporter from India revealed on *The Nightly Show* that (in secret and in relation to women) men think by "The Dick!" This anecdotal evidence confirms independently the hypothesis regarding the world being ruled wrongly as a violent whorehouse while lying and operating by hypocrisy. So, all the lies, conflict, racism, wars and sufferings going on currently are caused by egos lying to try to justify their confined in the darkness snaky Dick into a drunken woman or one in dark fear or one burning by envy.

We finally found humanity's central enemy and it's "The Dick!" It has only one purpose: maximizing the sum of: "up-in/out/in/out/...come-down/leave." Therefore from the prospective of a woman, the question isn't about what ends justify what means but rather what particular men's "means" justify their Dick's ends! The wrong conclusion that most attractive looking women must have reached unfortunately, in secret, is that looks, fun, control, fame and luxury are means that justify a Dick's ends.

To change the downward vicious (somewhat exiting to some) cycle leading to much suffering and death; of men pretending hypocritically to have a "philosophy" while, in secret, thinking by The Dick, at the center of the ego; and of women pretending to love while in secret acting like a slut, bitch or whore, both must correctly understand the correct and the right paths and ends; as they have been described.

I now justify all moderate women sluts whores and bitches who in the Independent Demilitarized Disarmed States will freely justify any man of little "looks and means" by allowing him— only the men in New Independent States and excluding me— to do safely what Doggs and Dicks rap publicly and popularly: "you already know, I wanna fuck you,.. you, I wanna fuck."

The needed change starts by "cutting" to expose "The Dick," by circumcision; physically for the Jews, and more importantly for all intellectually and spiritually. Then one needs to cleans and keep regularly cleansing their ego. Ladies, protest with banners saying: "be reasonable!"

"He guides me in paths of righteousness for his name's sake. Even though I walk through the valley of the shadow of death I will fear no evil, for you are with me." (Psalm 23.3-4.)

14. Women

All the statistics about violent crime everywhere in the world show that women are less criminal than men. Therefore, as a generalization— and as has been explained because all generalizations are descriptions of probabilistic wave forms and curves they have significant exceptions— women are less wrong than men.

Women are also the last, best and most beautifully balanced design of God. So, increasing the leadership roles of women in politics— not Hillary— and in business should be a global strategic priority.

There is a flaw in the design of humanity: Unless women learn, know, believe in and follow right, before knowing about evil and good, they, just as Eve, are bound to be getting attracted to and fn. Satan, while loving Adam.

As a result of being broken within themselves, like all hypocrites are, they and men are bound to be in shame, be in pain and die, i.e. get broken completely, spirit from body, within a Day, as God warned truthfully and as Satan lied deceitfully.

As the Old Testament shows, subsequently, when Adam and Eve died in less than 1000 years, and as Ap. Peter wrote in 2 Peter 3:8 and as in the Psalms 90:4, and as I showed mathematically, (considering resting time,) "one day for God, is like a thousand years" for humanity. Also, one day with God is like living a thousand human years.

Without knowing Right, one has no correct criterion by which to judge good from evil. As a result, one reverts into wrongly thinking that they are good and others are evil.

One must know, believe in and follow Right— from "the Tree of Life"— before knowing and judging about good and evil. Now you know.

Ladies, are you going to continue letting hundreds of millions of women remain enslaved as sex objects for fn. men, by violent, wh.house systems in hypocrisy, in the Middle East, around the world; and with much nicer methods admittedly within your own country?

Ladies, are you going to let each and every nation in the world whose competition wh.house system is designed by us men— in which whores who act like men can rule as well — be the only system and not demand at least a small Independent alternative, run by women mostly based on cooperation? A system that isn't violent, deceitful whoring as is that to which you have to send your daughters and sons currently?

Lady aren't you going to demand an alternative that is not based on force and punishing but is for those who choose non-violence and is demilitarized and disarmed, because it is you, who in relative terms are the weak physically whom the Right One, Jesus, is blessing?

And even if you like the current system, are you going to let hundreds of millions of other women who are or feel enslaved, abused, victimized and tormented remain into it, without giving them the option of any Independent safe alternative in this whole world?

The pacifists, the righteous, the poor, the depressed, the pure at heart, the disenfranchised, the tormented by wars, need, should have, can and must have an Independent Demilitarized Disarmed Democratic Land and State that is led by a Congress that is comprised by over 51% and up to 70% of women, and whose first President is a woman.

There is no man who is a racist against women of another race; almost all men of all races, if without repercussions, would gladly sleep with any pretty woman of any race. Racism, ethnicity, sexism, national wars, Islamic terrorism and other wars are about men, pimps, fighting.

Jesus said to them, "The sons of this age marry and are given in marriage, but those who are considered worthy to attain to that age and the resurrection from the dead, neither marry nor are given in marriage. (Luke 20:34, 35.)

Let me "declare it" to you i.e. explain it: The conventional interpretation is that people will choose to be celibate. Based on this interpretation there are Monasteries, including a whole Peninsula of Macedonia in Greece that has only monasteries, called Mount Athos that is autonomous, demilitarized, disarmed about which reporter Bob Simon said "there is no place on earth closer to heaven than Mt. Athos."
There are other many places where people have attempted to live independently and separately such as the Kibbutz, in Israel, Ashrams in India, Temples in Tibet and elsewhere and several Christian communities, like the Historic Peace Churches in the US. I liked all of those which I visited.

However, the fundamental disadvantage is that none of them are truly Independent of the State and as a result they become part of or get dominated by the power/control and money issues of the State.

For those who want to get married, to marry or are married, staying in the current societies is best.

Marriage and family are the central institutions that are the shields from the wrongful cultures of this world where a child may experience true love and where a family member may experience true love.

Having divorced twice I am not qualified to write this but I am told to avoid divorces because it is a whorehouse instrument.
Try to learn to tolerate, endure, the bastard and the bitch.
So within these systems, as Socrates said, "By all means go ahead and marry; if you get a good wife, you'll become happy; if you get a bad one, you will become a philosopher." That's how I became a philosopher. Then, I even became a philosopher[2] i.e. a philosopher of philosophers.

The institution of marriage must be strengthened for these current militarized competitive societies to work much better.
The way to do it is to "kick" many of the singles that would prefer staying singles and have whatever adult heterosexual sexual relationships they choose, without getting married, into the Independent Demilitarized Disarmed Democratic State.

Preferably people in these Holy Lands will choose lifetime relationships but by not marrying they avoid the potential of marrying for money or power, which happens as you might suspect in the current societies or not chasing power and money to get someone to marry you, which also happens and thus remove the last instrument for whoring, to be sure that the spiritually Resurrected who will move into the small Independent Demilitarized Disarmed Democratic Holy Land and State, within the territory of your nation, are not Whorehouses.

So, even though I find nothing wrong with the interpretation of "not marrying" to mean celibacy I believe that the interpretation of "not-marrying" to mean just not-marrying and doesn't have to do with whether these people choose to be (physically) celibate or be sexually active, is also valid.

As with everything else, except for God who is Absolute, marriage is a relative thing—Einstein named his theories relativity to emphasize that— and can be good and/or bad; in the context of defending a family and children from a deceitful, hypocritical violent government system and culture it is very good, if it can last.

Marriage as discussed by Ap. Paul was intended as instrument of Justice, to reduce abuse against women by men who by force, money, fame or other means had sex and then disrespected women and did not treat them as equals or left and didn't perform their responsibilities as fathers. Marriage can still do that.

Removing some of the singles will help marriages last within the current militarized societies.

However in the context of not using sex as an instrument for survival in the new, (derived from the Greek Neo,) Independent Demilitarized Disarmed Democratic States, marriage is not good.

There should be no need for any legal paper to confirm or deny love. The need itself for a legal or religious paper for marriage is more evidence of the absence rather than of the presence of true love.
One may generalize this into: the more legalistic any society becomes the more evidence there is that the intent and reasoning is still dominated by wrong.
As a result of being wronged a lot, people are forced into resorting to Justice and the Laws even if consequently they end up in a Labyrinth.

Not marrying and not wanting to get married (or to get married again after a divorce) is better for some, (including for me.)

Letting the few single ones who choose and commit to live a physically non-violent life and specifically choose to pursue lives that do not include controlling others, having much more, if any more than others and not trying to lure others by appearances for sex nor rejecting sex because of appearances, live free by cooperation, in true Peace and Joy in one Independent, verifiably Demilitarized, Disarmed Democratic State will help greatly both them and the current militarized competitive society.

From my point of view, all the issues of morality regarding sexuality aren't about the sexuality but are about the intent. If the intent or methods are not power-force, money or superficial appearances by a system designed to minimize those, it would be wiser to let women choose what sex is right for them and what isn't and not impose men's rules on it.

Sex was one of God's best inventions. Whatever problems of life one isn't capable of processing and releasing intellectually they can release them physically through sex enjoyably.

It is much better though to become Aware of and solve one's problems, including one's emotional and physical problems, intellectually.

That is why fn. time by my old age, the options available, and my Lord have conspired and put me under involuntary chastity for over 10 years!
The last time I went to a bar I came out with 0 women, 0 women said hello, and 0 women looked at me— maybe because I didn't look at anyone either, given that I talk with an accent and look and act like an alien since birth no matter

which country I am in— which convinced me that being an Alien Ellin Greek is tragic!

Even though God may have designed tragic Greeks for your entertainment I feel that it's sort of unfair and something has to be done to change it.
The cause of tragedy is not-knowing the end.

If Odysseus had read the Odyssey and knew that going to war would result in him having to go through hell and high water, and then even if victorious and even if he was able to get back to Ithaca, through more hell and more high water, he would find that Ithaca— a beautiful island but one that is and — was an intellectual dump with hundreds of men trying to f. his wife, as soon as he was asked to join the war against Troy he would undoubtedly and without hesitation say:
"F. you and fuck your ego Agamemnon. Fucking is less bad than fighting.
Do yourself a favor and let Ellen stay in Troy and send your wife and her lover to the Trojans because your real and worst enemy is within you and within your house. I declare Ithaca Independent of Greece," and Odysseus could have stayed happy in bed with loyal Penelope!

Also Agamemnon would not end up getting murdered by his wife and would have understood what Buddha later said:
"The best of conquerors is he who conquers himself."(Dhammapada 103)
And what Confucius said: "A gentle person blames self; a vulgar blames others."

Because he didn't do so, poor Odysseus after going through all that hell and high water he had to also kill much of the male upper class of his own island because it was they who were trying to take over his kingdom and f. his wife.
And despite his epic victories, when she met him again even Penelope was skeptical, not cynical though, and asked him a trick question.

Poor king Odysseus, unlike you, hadn't read the Odyssey, he hadn't even heard of it, so he had to go through lots of tragedy, epic suffering and pain— o.k. Calypso along with her nymphs may have been a bit fun— as all materialists have to because the material end is inevitable tragic death.

If poor Odysseus had read in the Odyssey how much damage to others he was really capable of and the huge price that it carried and had also read this book, when asked to join the war he would undoubtedly and without hesitation have sent a messenger saying:
"F. you and double fuck your ego Agamemnon; I declare Ithaca, an Independent Demilitarized Disarmed Democracy. Here, I have stuck my double edged sword on a rock for only an excelling Guardian to pull out. You better sign this non-aggression Treaty or else you will have the experience of a naked pretty looking virgin caught up in the orgy of a barbaric battlefield. And then, not even the brave Achilles will be able to save you! Sign..."

But then Homer wouldn't have had anything to write about and all would still be Barbarians and would not even know it. It is a good thing that now at least some barbarians are figuring it out because it is the only way to change it!
Ha! Ha! As one of Germany's Germs would dictatorially say: If this book has not made you at least laugh a lot yet, you flunk; you must go back to the beginning of the book and start reading it again!

Do you want to become a war veteran and after going through hell and high water for years come home to find out that home has become a dump and hundreds of men are trying to f. your wife? Learn, my friend, learn right, learn it now or else tragedy will strike.

Are you willing to take moral responsibility for each of the unnecessary deaths of all those innocent children that are being bombed and killed in secret by drones and are called "regrettable casualties?"

Friends, we don't need nor want any defenses because what we do is right. Don't allow fear to rob your lives; we need not and will not fear.
The Greek word for a person is "atomo" and means indivisible.
Right Reason, the capacity to reason rightly, the Christ—Logos— is and has been "within" each person all along and is the only one who can make you indivisible.

Do you know what you are doing while staying in these hypocritical, deceitful violent whorish systems? You are killing the Christ within you and are crucifying the Christ amongst us. Now you know what you do.
And there is no excuse left.

The only way to avoid the tragedy and see the comedy is to rise beyond the solid material and into the air like, spiritual.

Gain Knowledge from and of our God, Elli, the True God and *let go of fear; let go of possessiveness and pain; let go of envy, greed and shame.*
These arise from the ego so let go of the ego and…Rise to the Heavens to join The Christ Jesus and us.

Forgive me for being Greek and making you uncomfortable but have the people learned anything from those that they say they believe in?
One cannot be correct ethically without understanding the morally right that they must be consistent with.
Did we provide the morally right — and in the context of only bad options— the ethically correct education?

Please re-think, to reinforce the synapses in your brain, what you learned about right and about correct, i.e. consistent with right. So, that you may change the current truth you are experiencing into being one or more of the Eternal Truths.

It is extremely unlikely for one to be right in the wrong context because one wouldn't have been in the wrong context if one was right.

One cannot be morally right in the context of the, as I have proven, wrong systems, even in the least wrong systems that are designed to reduce wrong, without suffering for it greatly being in poverty, weak, being taken advantage of, create co-dependencies, get abused and be in mourning. Jesus Christ could not do it and survive long, what arrogance makes you think that you can?

If one wants to be right they must move into the right system that is by design non-violent, non-threatening to any, supportive of those who stay behind, free and not-whorish, as has been advocated.

If one stays in these systems the best one can do is to be ethical, to be correct, to admit upfront that they are in part wrong and to be least wrong and consistent with right, with ethical standards (that are sorely lacking these days) which must be much higher than the law.

The exodus of some into one Independent demilitarized disarmed State will greatly improve the morality and well-being of those who move there. It will also greatly increase the standard of ethics within the current militarized systems in which self-righteous deceivers currently only hypocritically talk about love, peace, truth, freedom, forgiving, harmony and righteousness but have no experience of any of them.

Presuming people are righteous and thus are clean and safe in their intentions, reasoning, emotions and body and are in the right context i.e. in the new Independent Demilitarized Disarmed Democratic society that is advocated, they may choose to be celibate but as an inventor I see minimal problems with God's brilliant and beautiful invention of sex, which allows others to pleasurably experience one's problems, (while using Trojans as shields!)

I know that I am very likely to get accused for promoting fornication by "moral" hypocrites who worry more about what the vaginas do rather than about the spirit and intent.

Sex is one of the most private subjects and is least rationally discussed. I know and agree that in the context of the current systems, fornication that is the tool that is used to gain influence/prestige/control and/or to get money and/or to cheat others to gratify self is appropriately treated as a sin.

But in my opinion, that on this subject has no biblical, scientific or experiential evidence, single pretty women who have clean, safe sex for free with the not-good looking, the poor, the weak and the depressed in the correct context, particularly now that the Trojans have become condoms, are doing sorely needed charity work and should be honored.

For those who stay in the current hypocritical money-power systems, 40 days after their relationship with someone good-looking or rich or strong or "happy" breaks-up I would even go as far as to advise that they go to pilgrimage to an Independent Democratic Demilitarized Disarmed State and fuck physically for free a not-good looking, poor, weak or sad adult of the opposing sex to atone for their wrong choice, injustice, sin and to recover emotionally.

If anyone condemns any woman for doing so tell them that Apostle Pol approves it and if that doesn't work, tell them that your bulky bitchy friend Paul tells them to fuck off.

It doesn't have to be so; being Democratic people may choose to marry or some married people may want to live there but some statistical separation will help both sides.

From my point of view, the women who have physical sex, not for the dirty reasons of power, money, appearance, lust, fame for themselves but for their love of others and because it is right for the other in the right context, are and should be called (if they want to,) Virgins.

Like the Virgins of the Parthenon, that means a place for Virgins. Parthenon is derived from "parthena" that means virgin.

The Golden Age of Athens was not driven mostly by the men of Ancient Greece but rather by those Mystic Virgins in Spirit women that spiritually held up the Parthenon.

We don't want the statues of the Virgins that held up the Parthenon back; fk. the stolen from us statutes as far as I am concerned.

We want our pure in heart, honest, truly correctly loving Virgins in Spirit back, who have been stolen from us and are enslaved in violent hypocritical systems by whorish Babylonians and violent Barbarians who pretend to have gotten civilized… by us.

We desperately need those blessed pure in heart (Matthew 5:8) Virgins in Spirit to hold up the New Parthenon and the New Jerusalem in one Independent Demilitarized Disarmed Democratic Holy State.

Ladies, Virgins, Resurrect. Resurrect and testify about who is Righteousness, what is right and what correct; and testify please about who is correct in telling the Truth, the whole truth and nothing but the truth.

Don't forget that "the student is not greater than the Teacher," as Jesus said. (Matthew 10:24)
Which people taught you everything that you know?

We will have to get all to resurrect to resurrect the honest fearless spirit of the conscience of the Virgin and to resurrect her/his right reason.

Consciousness is the recognition, the experience and the reaction to truth; to the external and the internal facts.

Conscience is the consciousness of other people's consciousness.
Resurrect; this is the Day of Judgment of the very, very few and of Deliverance into true Liberty for most.
Some Ladies have been chosen from the Beginning to be truly free, Independent and in peace. Don't dare to stand in their way any longer.
I do spiritual and intellectual war to avoid imminent physical war.

Because all I do is war the best role for me in the future is to do nothing and actually retire as I was intending a quarter century ago before I was informed that God had a different plan. If I am invited though to any Independent Demilitarized Disarmed Democratic State, anywhere in the world, I would gladly go and do nothing there. I hope that there will be a place for those like me in the US and that I will not have to immigrate again for a 3rd time because immigrating is gut wrenching and part of the 'old self' gets to at least temporarily die. I unfortunately know how my forefathers, Greeks in the diaspora, felt.

Ladies, get out of the violent militant systems that fn. men invented and mostly run, to reduce wrong. If you like your current state, at least help those who want to get out of the hypocritical whorehouse systems and the rules that pimps, who failed to inspire their spirit and to teach their reason rightly, have imposed on their vaginas, get out. Govern or let them govern rightly and run independently and safely the non-militant non-violent societies, as you/they believe is right.

The biggest risk for the Independent Demilitarized Disarmed Lands is that they may become too hedonistic. Remember that the straights are and will remain narrow, even though their price for failure will be much smaller.

To make the correct judgments about good and bad not only you must know right, first, but you must also "know self" i.e. know the truth about yourself and the correct balances within your spirit and body. The balance between meeting self-interest vs. the interests of others is and will remain narrow and meeting both requires honest straight correct reasoning and dialogue upfront, before any action; and must be getting rebalanced correctly, regularly.

Conservatives dislike change. For example, Hannity asked: "can someone love you if they want to fundamentally change you?" If you are self-damaging then one does love you if they want to change you. If someone is wrong they are, whether they experience it at the time or not, self-damaging.

Alexander the Great when dying was asked who he declares as his successor, didn't appoint any of his Generals but said "whoever excels." The narrow honest, straight, less travelled path requires, demands, excellence. "Aristo," means excellent. Aristotle taught that virtue is the by practice pursuit of morally right, intellectual and emotional excellence.

Odysseus had to prove his excellence as an archer by getting his arrow clear through the center of 12 double- axes, like my arrow through the 12 Crosses of the 12 Academic Disciplines.

Then, Odysseus, with the help of his son whose name is Telewarrior, killed the Suitors who were led by a strong man named Irrational.

"Wherever there is Chrishna, the master of all Mystics, and wherever there is Aruna, the supreme archer, there will also certainly be opulence, victory, extraordinary power, and morality." (The Gita; Text 78)

The most critical factor is the motive, the intent, that you have to discern by seeing spiritually as you now can, about you and others; is it love of others but without getting significantly harmed which is correct; or is it love of self supposedly "without harming others" which is incorrect because there is harm, serious moral harm, ultimately deadly harm, in not truly and correctly loving any non-kin other?

Loving rightly and truly cannot be done without exceedingly significant unnecessary suffering in this world's systems, so that it needs to be done in a New Independent Demilitarized Disarmed Democratic Society.

"Put in your sickle and reap, for the hour to reap has come, because the harvest of the earth is ripe." (Rev. 14:15)

Each component of Satan's deceiving troika— that had been ruling this world— of hysteric (bitch) beast; borderline schizophrenic (slut) dragon; and narcissist Babylon (whore) is in direct opposition to each Aspect of God's Holy Trinity.

-The hungry for power and control hysteric snake, wolf, Beast rules this world by violence, punishments, fear through Might and is opposed directly, and contrary to what he says, he rejects the rule by the Right One, Christ Jesus and abuses and murders the righteous.

-The deceitful, lying, hypocritical, borderline schizophrenic, angry, fiery hot, short sighted slut Entertainment dragon is opposed directly to the Absolute Truth, the Holy Spirit, and to even the existence of truth making up whatever truth he likes for his benefit and calling it truth, grabbing all the attention and causing drama, envy and confusion.

- The money possessions possessiveness thirsty, ashamed, envious, greedy, confused, narcissistic whore Babylon abuses the love of others, wanting unlimited love from others while abusing them financially, uses people as things for labor, damages Love and is opposed directly even to the existence of Love, the Father.

Is this the right time to stop being ruled by Satan's troika of the Whorehouse in which sinful whores, bitches/pimps, jerks/sluts pretend to be virgins, Christians, Jews, Muslims or Buddhists resulting in suffering and dying, and to start being free in eternal joy by living by God's Right and True Love? What excuse is there for not declaring Independence for a small Demilitarized Disarmed State of Peace, while your government is taking people to war, while there is so much injustice that is rewarded, so much sin and suffering, damage to children, unnecessary deaths and so much unnecessary anguish?

Such action will be the best not just for those who voluntarily choose to be in the Great Exodus into one small Independent Demilitarized Disarmed Democratic Holy State, governed mostly by women but will also do most to help, with most benefits and least costs those remaining in the current militarized Democracies of our nations become more competitive and better balanced in defending all those Freedoms, sometimes heroically, who should be honored and truly helped, after the codependency is broken, for doing so.

"This is the judgment, that the Light has come into the world, and men loved the darkness rather than the Light, for their deeds were evil," says Jesus in John 3.19. I hope that this stops being true, now.

Once and only once one is in Oneness within themselves by the Oneness of the Holy Trinity and in Oneness, unity and uniqueness by mutual agreement among themselves, in an Independent Demilitarized, Disarmed, Democratic non-violent Holy, separate, State, there will be no Cross for one, and each will live forever with God in peace, liberty and Joy.

15. Is this the right time for the Resurrection?

How much more suffering, unnecessary deaths, oppression, diseases, disorders, disasters, torment and destruction do people and their current leaders want before they decide to do what is right? I don't know.

In the original text of the New Testament, Satan is called "ponoiros" which means cunning and causing pain.

I dislike pain both giving it and receiving it. In my view, there has been much more than enough suffering, anguish and pain already.

I know that people will either learn to think and act rightly or die as a relief from the torment they caused to themselves.

So, at some point, some will start reasoning rightly and establish at least one small Independent Demilitarized Disarmed Democratic Holy State in this world.

That Exodus— the great migration that is prophesied also in Islam— of the Resurrected and their children, the poor, lonely, depressed, disenfranchised, the billions of weak people living in quiet desperation, the pure in heart, the single parents and their children, the pacifists, the merciful and the righteous, whom Christ Jesus blessed, may live rightly, in dignity, in true Freedom, in lasting peace and in true and lasting joy.
I hope and pray that that right time is now.

All the profane, foul, and explicit language that was used was not at all gratuitous but was fully necessary to describe minimally the profanity, ugliness, dirtiness, sickness, whorishness, violence, cruelty, and deceitful, disgusting, devilishness and evil that is dominating this world.

I would need a few more pages of maximally profane and vulgar cussing to fully describe the profanity, barbarism, dirt, violence, brutality, oppression, whorishness, murderousness and hypocrisy of the super-dirty beasts that you call leaders of this world.

What is right for each to do is to start passionately calling for, in Social media and to their Congress person, an Independent Demilitarized Disarmed Democratic State for the poor, those living in quiet desperation, the weak, the disenfranchised, the pacifists, the righteous and the true believers within their nation.

Until the day that this is accomplished any contact with any adult depresses me.

Shouldn't every democracy have the option of: "exit this system" in each of their ballots? Is our freedom limited to which mentally disordered whore, bitch or slut we prefer?

From my prospective the spiritual, intellectual, emotional and physical suffering caused by very few humans at the top, driven by narcissism and obsessive need to dominate all others, without letting anyone out of their system, to the billions of the rest of humans and to all other species is immense and anyone looking, participating and interacting with it and not getting depressed or angry is either spiritually dead, a zombie or too fd. up.

I don't want to learn any more about evil; I learned enough. Until that day, the day that such an Independent Demilitarized Disarmed Democratic State exists, every day that no adult, other than my children, contacts me, is a great day.

Lao Tzu wrote: "If the Wise is compassionate why has he disappeared into the Wilderness? The Wise being in harmony with the Tao is invisible in the world of greed and among those who hunger for power. How do I know? Silence tells me."

If you want to do right and good show it; I told you how.

Those who stand in the way by action or inaction of the enslaved in hellish violent whorehouses to be free and to find their dignity, love, liberty, peace and joy are evil hypocrites and in my judgment should burn in hell.

Does the current unrighteousness have to get to the point of World War III breaking out, blowing up the earth and making most of it uninhabitable for decades if not centuries, for your leaders to do what is right?

Does this "me generation" have to be evil and unrighteous like Jesus said about His generation and let this change that is right, should, needs to and can happen, now, happen after I die from this world and ascend?

I have "exited" this system already long ago and have recently found joy. I found joy and peace, even in isolation. (Some men from an island are an island!)
It is the joy of **being** that exists within each from early childhood but it is taken away by not giving people enough time to be.
Bliss feels like the feeling after a victory having accomplished what you were supposed to. Bliss is like the very relaxed and satisfied feeling, right after sex, only it keeps lasting. It's like: 'I just gave a fk., so I no longer give a fk.'… about how the rest of the Universe is doing.

Physical health
Daily enjoyment is found on little things like a smile, physical non-sexual affection, a sport, dancing on an empty beach at night, a hobby, nature, and even in the most necessary, most urgent, most cleansing function of our bodies which is breathing.

The politicians of California (even though we have an ocean of water next to us) currently suffer from a severe drought in their thinking because they lack living water. As a result they are forcing me to not water enough the trees in my garden. To clear my mind from such external interferences I do yoga while repeating my mantra that's: I don't give a fuck!

While this probably the correct mantra for us ignorant sluts/jerks, the correct mantra for the immoral, incompetent whores/bitches/pimps and politicians, is: I **give** a fuck. Breathing rightly, by taking a deep breath while expanding your tummy and breathing out fully while contracting your tummy (as taught by yoga,) while stretching the areas that hurt and while repeating: I give a fuck, helps most the body!

"Whenever there is a decline in righteousness and an upsurge in unrighteousness, then I manifest Myself (in personal form)" Gita 4 – 7.

"Very truly I tell you, no one can see the kingdom of God unless they are born again." John 3:3

Only in John's Gospel, The Christ Jesus says: "Amen, Amen," which means "the Truth of Truths" (or "the truth about the truths") and it is often badly translated as "very truly I tell you" or "I tell you the truth." Both are bad translations because they imply that the rest of what Jesus said are either only a "little truly" or lies.
You need to be born again, not in body this time but in Spirit to See the spiritual kingdom of God.

So, given that there was a software glitch in people's mind someone had to fk. their old wrong thoughts and bad spirits, causing you all your troubles, for you to get the new right, truly true and loving ideas in which to be reborn. I am sorry it had to be me but someone had to do it!
I still respect you very much.

It is come to Jesus time...

The only way to overcome the conflicts by the inevitable oppositions in life is by loving the enemy, by loving the opposition.

The Greek Mystics were expecting the Christ just as the Jews were expecting the Messiah. Many claimed to be Christ or the Messiah.

The only One who taught, lived and died by loving even His enemies is Jesus which is one of the many reasons but a sufficient reason as to why He is the Messiah, the Christ.

There is no way that any regular human could have figured to love their enemies as the answer to humanity's ills and no human would be willing to knowingly go through so much suffering to prove it, as He did.

John 1.1. "In the beginning there was ΛΟΓΟΣ." This Ellenic word was central to Ellenic philosophy, is transliterated as LOGOS and it means Right Logic, Right Reason, Right Reasoning, Right Purpose which is expressed through right words i.e. the "Word" as most English translations translate it.

"But I tell you the truth, it is to your advantage that I go away; for if I do not go away, the Helper —an exception of a name translated; the Ellenic word is "Parakletos" and some translate it as Paraclete and others as the Advocate and others as Comforter— will not come to you; but if I go, I will send Him to you. And He, when He comes, will convict —criticize, judge, prove wrong— the world concerning sin and righteousness and judgment; 9concerning sin, because they do not believe in Me 10and concerning righteousness, because I go to the Father and you no longer see Me; 11and concerning judgment, because the ruler of this world has been condemned." (John 16: 7-11)

Is this passage applicable today?

Righteousness is more that Right Reasoning and includes right intent, right understanding of the context, right attitude, right prospective, right purpose, right objectives, right strategy/concentration/focus, right goals, right priorities, right relationships, right words, right actions, right measuring criteria and right results.

When it comes to Buddhism I don't understand the original text as I do the New Testament and the Ellin philosophers, so I have to go by translations. Therefore, I cannot say for sure whether it's correct or not but at least an interesting and valuable, if not right or correct definition of the components of Righteousness, is by Buddha, as the eightfold Path to Enlightenment.

The way I understand the components of the Path are: right understanding, reasoning; right intent, attitude, prospective; right speech, expression; right action; right livelihood; right effort, energy; right awareness, of the spiritual reality and of the facts, mindfulness; right concentration, focus, "wholeness," oneness, enlightenment.

Specifically in John 16:10 Jesus calls Himself Righteousness. "Concerning **Righteousness**, because **I** am going to the father and you will no longer See Me."

One of the many other reasons that I am so sure about Jesus Christ being Right is that there was a time that I very much doubted Him. I was like Ap. Thomas—a name of Aramaic origin from Syria— who must have been a skeptic i.e. doubting everything and everyone.

The reason I doubted my Lord and needed so much evidence—Jesus blessed those who believed without having needed all this evidence—was that despite living in several different countries that pride themselves as Christian, their cultures were surely screwed up.

So, because all problems can be traced to the top, for a while I considered that maybe Jesus said or did something wrong and I looked for something that could be wrong or some inconsistency between His words and His actions and because I was a slut, I also looked for some better teaching in another religion.

I found no better teaching and found several flaws and inconsistencies with what everyone else has said and did, but none in Jesus.

I was helped by M. Gandhi's correct statement: "I would be Christian if it were not for Christians," And: 'I like your **Christ**, I do not like your **Christians**. Your **Christians** are so unlike your **Christ**.'

It is then that I realized that we have all been wrong, because we have been unnecessarily forced by extremely wrong, ignorant, frightened, selfish, egotistical, spiritually blind leaders through militant punishing survivalist whorish systems into being wrong and into being hypocrites by not admitting it, and that only Christ Jesus is Right.

Jesus is Right and is Righteousness as evidenced by what He said, how and for whose benefit He reasoned, what He did, how He Lived, what He intended, why and how He Died and why and how despite having died in body, He now lives again and has more impact on people today and everywhere in the world than all the dead-living combined.

Hypocrites cannot be forgiven because they pretend to be right and do not admit to having been wrong that is a prerequisite to changing.

Virtually all modern scholars of antiquity agree that Jesus existed historically. It has been evidenced and documented independently by many who had no reason to lie about the virtues of someone else and documented even by non-Christian historians.

I agree with everything Christ Jesus says and does as recorded in the Bible. That "agreement about everything," as per the original of Matthew 18.19, liberates from everything, makes anything possible and results in both Oneness and in uniqueness.

Truly believe in Him, Righteousness, believe in right, intend right, think rightly, feel rightly and most importantly do right and live right. It produces the right for you results.

Christ Jesus is The Light that Enlightens.

Logos, the morally Right Reasoning, The Christ, the divine capacity to reason rightly, has been Gifted to every human and has been within you all along.

Christ Jesus is your righteous conscience, who was guiding you throughout your childhood, until you were pretty much forced by a whole system to put it aside and silenced Him, in search of vain power, control, fame, money, entertainment and sex, in the name of survival.

Accept Christ Jesus, Logos, right reasoning, right "Word," righteousness, within you and make Him your Lord and Savior.

If you understood what has been written your conscience has been resurrected and your capacity to reason rightly has been restored.

I say: This world is wrong and people sin because we didn't truly believe in righteousness enough to be doing what is right. And this world is wrong because we didn't see righteousness until now in this world, because the Christ is no longer with us in body but only in Spirit. So, that the right judgment is that the ruler of this world is extremely wrong and is the liar Satan ruling a hypocritical Whorehouse with force and violence through the leaders of the nations, and he has been condemned.

Didn't I just do what Christ Jesus said I will do?
Didn't Jesus Christ's prophecy about the Advocate come true?

Isn't the proof of someone being right, in the context of time, that His prophecy becomes true?

There is no personal benefit to me to claim that someone else is Right. As one of the honorable Greeks because of the few Great Ellins who brought civilization to the whole world whose names you know, who still live amongst us in spirit, there is no benefit to my nation or people, to testify and provide independent evidence that someone else, none of us, and particularly a poor, weak, tormented Jew, Jesus, was, is and will be Right.

What have the false prophets and scribes, who intellectually support the spiritually blind, extremely wrong hypocritical political leaders who lead these violent deceitful whorehouses prophesied correctly and with what evidence do they claim to be right? They are dead-living zombies living in body but are dead in Spirit because their conscience has died.

The Source of Life and Judge of all Creation, the Holy Spirit, says that G.W. Bush, W. Clinton, B. Obama, V. Putin, A. Merkel, Ayatollah Khomeini, Kim Jon Un, Xi Jing Ping, and most Presidents of most nations these days have been and are very to extremely wrong and are hypocrites in pretending or claiming to be right or even to being righteous and they should burn in hell unless they admit to have been very wrong, become righteous and prove it by establishing and recognizing the Independence of one small Independent Demilitarized Disarmed Democratic State in their nation or at least one in this world.

Even though they claim differently the economic leaders of the world, the "Intelligentsia" as well as the political heads of the nations, are trying to concentrate power to them-selves and ultimately rule this world through a world government, because of increased "inter-dependence" they say, a codependence they create but which if you allow you would be too stupid.

Are these leaders looking to empower, enrich and support your happiness or are they looking for their own and their Party's power and wealth? They are all very wrong, sometimes extremely wrong and are hypocrites by not admitting it.

I don't claim to be right and do not believe that anyone other than those blessed by Jesus Christ is righteous; those who pretend to be righteous are as I have shown hypocrites. As to whether I am righteous it is, by definition, up to others to judge.

I claim that this book is the least inconsistent teaching of everything, every major academic discipline, everyone, and of God, with Right, Jesus, and by that criterion it is in my judgment The Least Wrong New Book.
Did it reveal the Revelation for you?

I should have entitled it or subtitled it: THE BEST NEW BOOK.
That's it; I have had it with my big bitching ego also. Didn't Jesus say: "Do not judge so that you will not be judged. For in the way you judge, you will be judged." (Matthew 7:1.)

If a human is to judge rightly they must judge straight, like with a rod of iron. How can someone know and judge who the hypocritical very big bitches, pimps, whores, jerks and sluts are without having been one of them? My big bitch, pimp, jerk, slut, whore ego is wrong, is guilty and must be condemned. Shouldn't I kill my ego?

The hypocrite was doing business with nice people and while not using any vulgar words and talking nicely and politely, in meeting after meeting he kept thinking about those nice people: who is the too greedy confusing whore; who the scared, paranoid over extended, over controlling bitch; who the too short-sighted angry slut; in what combinations; by how much; and who is Dead; intolerable.

What is worse is that he would then try to hire the biggest spiritual whores he could tolerate into Marketing, Engineering or General Management and encouraged them to be more greedy, for growth; the biggest sluts into PR, HR, Sales and Service and incented them to be more short sighted and look for quicker gratification, for improved current results; and sought out the most paranoid and controlling bitches he could tolerate into Finance and Operations to control the other two groups, for higher ethical and quality standards and encouraged them to be more paranoid, to bitch more and to be and more controlling; Intolerable. Unfortunately, he was repeatedly successful with it.

And then he would do team building activities to get the executive big whores, bitches and sluts to like each other, inform them that the less they ask questions and bother him the better they are doing and then he would leave in the middle of the day to go fornicate; intolerable.

That is why he knew; he was not just theorizing but also knew from experience that the biggest offenders who hypocritically deny being the biggest offenders are the ones at the top of the political and economic hierarchy.

He saw those "leaders" operate. He taught hundreds of CEOs. He knew exactly what they are doing and he knew that their actions are very wrong and are in direct opposition to right and to everything these blind "leaders" say they believe in but he kept being "their friend."

If he had to do it again in this world he would probably run corporations this same way again —with more non-executive employee ownership and more democratic this time— like growing an ethical spiritual wh.house. Instead of regretfully admitting his wrong doing, by acting fancy cute and classy he hypocritically pretended that he was moral, which makes it all much worse.

No wonder he felt bad and guilty; he was very wrong, even if he was least wrong under the circumstances. He must get killed. He is the world conqueror king of the Greek Macedonian— Mache means war —war heathen. There were none who excelled so it's Alexander the Great's Empire that's now again on fire.

He is not kidding in saying that he is ready to get W. War III going. The balance of power is moving rapidly East and if what is suggested in this book does not happen the power shift that seems smooth now, will become sudden in a couple of years sinking the West into another recession or Depression while China starts its dominance.

He comes from the East and is so bad that he is willing to, (and if I let him he can, because too many Greeks are already fed up out of their minds,) take Greece not just out of the Eurozone but also out of the EU and even out of NATO, if things don't change fast, and align Greece with China, Russia and Iran giving them, not for nothing, military bases in Greece and thus in Europe, with direct military access to the Middle East and Africa also making that shift a lot more severe, pulling Southern Europe with them and either sinking the West into a Dark Age again or starting WWIII.

So please solve the problems as advocated above rather than keep making them worse as has been happening until now.

And he was wrong in asking God to kill him because if God killed He would not be absolutely Good. I should kill my beastly ego.

Oulala; Am I supposed to feel scared now? Why don't you tell me what scared is supposed to feel like, so that I don't disappoint you Michael?

In war for Victory I make art my business also. Can a Charioteer go to battle without synchronizing all his horses by artistic beauty in his chosen direction...? In fact, you can't do anything well without at least some art and you admitted that you are clueless about art. Do you know why the head of Victory isn't on the Statue of the Victory (from Samos) in the entrance of the Louvre in Paris? Do you know where Victory's head is? I know that you are an Archangel and stuff but being clueless about all these things do you really still want to challenge me, Michael?

Remember, he called this whole world not just wrong, which it is, but also an oppressive hypocritical whorehouse with violent outbursts, which must have offended everyone.

I have admitted publicly my wrongs and sins. Have you heard of Due Process of law? Or is that too English for you? I want my lawyer. I did find a flaw in Jesus' teaching and it was that the Paraclete hadn't fully done his job. Whose fault's that? And He couldn't have done his job without me and my profanity. Where is the fn. thanksgiving for that? Also, the program Smith is out of control because the resonance of the third harmonic is way off so you will need a much bigger than 1024 matrix to deal with that.

He is in trouble now and he knows it; whenever he was in trouble he would confuse people and make them feel stupid by artfully throwing technical terms that most wouldn't understand.

Now you can see the whole truth: the Truth about the Truths and the truth about the lies.

There is no room in this world for a beast, like a proud white horse insisting in victory. Nor is there room in this civilized world for an African lion commanding and demanding his territory.

Don't worry; Neo will not hurt much. I have a double edged sword that is so sharp and fast that it shortens time, so that he will barely feel it.

I have killed my beastly ego before, as it is necessary for good parenting but he has the bad habit of waking up again when too disturbed. Boom; there; he is dead…

I feel much better already. There is no bitching anymore in my head. No one is sad and looking for something better. No one is feeding whorish greedy pigs anymore.

Lao Tzu is right in saying that "I suffer because of self. If I were selfless why would I suffer?"

He was a nice guy; he did the best he could. He was just a Man, what can you expect? He had big cojones. He was a hard working Ellin despite looking like a lazy Greek. He is forgiven. He did end-up writing a couple of funny things; we are thankful. Forgive his profane intrusions into this book. May he rest in Peace. Thank you for having attended my ego's funeral services. Have some coffee and a cookie…

Jesus Christ; the Buddha is also right in saying that the ego, the soul, is a temporary illusion. Being without the ego, without the old animal soul, is the Spiritual Deliverance into Enlightened Eternal Joy.

Confess that you have been wrong also and confess your past sins...cleanse your-self by asking for forgiveness... surrender your life into Righteousness, forgive self and others truly... so now you are reborn; born again in a New Spirit coming from God. My friend, you are now Resurrected.

You are no longer guilty because of your righteous faith and your testimony about Righteousness, Christ Jesus and because you didn't know that the systems of this world encourage wrong and sin and enslave into wrong bad options.

Billions of living in Spirit and in body people in all Continents, races nations and ethnicities and I, who were saved and forgiven by The Christ Jesus and who are now liberated by His Second Presence, provide independent overwhelming evidence by our faith and our testimony that He is Right, He is Righteousness, He is the Logos, the Logic, the cause, the Right Reasoning and the Purpose, the Son, with and through whom God Created and Creates the Universe. Can you See, can you visualize... and feel being truly and rightly (for you) loved by God and by many people by words and by actions and being in Joy, Truth, peace and free? You now See God's Kingdom.

Christ Jesus, YASUYE, Righteousness, is the king of kings, the Truth of Truths, The Amen Amen, the I am because I Am, of God's Kingdom.

Now you are free in Spirit, at Home in Heaven by one or more of the righteous Spirits, the Self-Evident Eternal Truths of the Holy Spirit, you reason rightly and are not enslaved into thinking wrongly and thus into sinning.

Now, it should be crystal clear to you that you should not partake in the worldly that encourage and then enslave into sin but should be separate, holy, and in a non-violent, New Independent Demilitarized Disarmed Democratic State that is for the blessed and the righteous who neither want to nor have any reason to be wrong or to sin by keeping the Spirit and the Letter of the 12 Commandments of The One True Living God, your Lord Elli.

I am eternally grateful to my children, wonderful Alexander and wonderful Emily for helping me learn about loving rightly and truly.
Alexander's contributions in helping me understand both the Unified Physics Theory and the Theory of All have been essential and critical.
The following is a nice link about Jesus, sent to me by my daughter Emily: http://m.youtube.com/watch?v=KGlx11BxF24

I have to confess that I don't have any great position; actually I am the least in God's Kingdom. It would be nice to have a big position in God's Kingdom; maybe I could impress someone who likes people with big positions!

However one prioritizes the Eternal Truths, I am at the bottom; they all look good to me. Being great in God's righteous Kingdom requires more ego sacrifice, and hardship than I can handle so I leave that to better people than me.

If it was up to me, I would have kicked my old ugly looking self out of Heaven long ago. I am not clear why they tolerate me in Heaven but they do and they love me. Also, they even like me when I don't speak about them but only about others, (as if I am too bitchy,) and they do what they can to help me stay quiet and happy!
I am at the Gates of Heaven and judge. That's most of what I do; use critical thinking that in its original means judgmental thinking to rightly judge.

Critical thinking is at the cross-point of the Cross —that you may imagine similar to the Crosses above— and is the result of the four functions of thinking i.e. analysis/ synthesis; induction/ deduction.

A righteous judge cannot judge by her/his likes or dislikes but by the law.

Even though forgiveness is better than judgment, if one has to judge, a righteous judge cannot be too forgiving of the criminals and/or sinners— without clear evidence of sustained corrected behavior — because that would be unjust to the victims.

So forgive the financial and spiritual debts that you hold on others, if you want your spiritual debts to be forgiven when you meet your Judge.

If someone is more screwed up than the very low bar, as you saw, of my dead ego they're too screwed up!
I judge by the righteousness of actions.

Rise and join Christ Jesus and us in Heaven; you are welcome in God's Heaven.
Now that you know the Secrets that need to stay Secrets from children because they are the ugly part of the current truth, you may call yourself a Mystic or an Ellin.

Please inform me at paulzecos@gmail.com where specifically you find significant inconsistency with the facts or with the Eternal Truths or with Christ Jesus' Words, Reasoning or with His Cross?
To maintain your integrity, your indivisibility, your oneness, forever, you need to understand the Oneness of the Trinity of God.
It is the Oneness of Love-Righteousness-Truth.

The One God, our Lord Elli is Loving Right and True.
The One God, our Lord, Elli, is the Absolute Good and Love, the Father; the Absolute Right Logic, Reason, "Logos," expressed as the Word, the Son, Righteousness, Christ Jesus;

and the True, the Spirit of Truth, the Holy Spirit, who is the Source of Life and the Judge of all Creation.

It is by that unity, that Oneness of the Holy Trinity, by being truly loving and merciful to those you disagree with and dislike, and by being truly righteous not just by words but more importantly by action and results and by being fully honest, true to your Higher Self and Just, that one becomes unique and undefeatable by any evil, unbreakable spiritually; a one, a "whole" and is eternal; a god, a Child of the One God.

Isn't that Oneness what Jesus prayed for?

"I do not ask on behalf of these alone, but for those also who believe in Me through their word; 21that they may all be one; even as You, Father, are in Me and I in You, that they also may be in Us, so that the world may believe that You sent Me. 22The glory which You have given Me I have given to them, that they may be one, just as We are one." (John 17:20-23.)

It is by living not by judgment under the law as is necessary in these worldly systems; but by living by love's oneness "within" through the Eternal Truths of God's Kingdom, along with the oneness by mutual agreement with others in a separate mercy based non-violent Demilitarized Disarmed Democratic society in an Independent New State that one may live forever in peace, liberty and joy.

"In my Father's House there are many rooms" John 14.2. I will repeat somewhat poetically, with some of poetic liberty, the Eternal Truths any of which one must make true in one's life to live eternally:

Love is **Righteous** and saves others, Love is True, IS, and is the cause and purpose of existence, love is eternal and others resurrects;

love is unique, creative and innovates, love is in oneness and Unifies (Christ Jesus);
love is kind and **Enlightens**, love is Free and others Liberates, love is competent and leads, love is joyful and delights (Buddha);
love is **Graceful** and harmonizes with its undeserving opposites, love is thankful for the opportunity to love, love is humble and others elevates, love is patient, persists and despite failures perseveres (The Vedas);
love is **Holy**, is separate and the holy from the unholy separates, love is discerning and distinguishes right from wrong and truth from lies, love is pure and cleanses, love is moderate and does not cause envy nor has envy (Lao Tzu);
love is merciful and **Forgives**, love is in peace and makes peace, love is hopeful and inspires, love is charitable and the needy helps, (Koran);
love is compassionate and supports, love is **Wise** and rightly teaches, love is courageous and conquers, love is Sovereign and self-governs (Confucius);
love is honest, truthful and integrity restores, love rightly understands, impregnates and Knows, love respects others and other life, love is **Just,** lawful, and the unjustly injured redeems, love is faithful and endures.(Moses and the prophets)

Jesus said: "And other sheep I have which are not of this fold; them also I must bring, and they will hear My voice; and there will be one flock and one shepherd." (John 10:16).

If one is faithful and true to any three—as their intent; their rational; and their expression— of the Eternal Truths, is there anyone righteous or Just who would find that one evil, wrong or condemn them?

I have proven that Christ Jesus is Righteousness, as He says, and that all seven righteous great religions are, as they each and all say, paths of and to Righteousness. Therefore: **The true believers of all the denominations of all the seven righteous religions are Christ Jesus' flock**.

Whoever disagrees with the statement above is wrong. Because, by definition, they either are ignorant and do not know what the founding documents of their own religion say nor know what science has actually proven — whether they believe it or not— nor know them-selves; or they are irrational and do not understand right nor believe in right; or they do not believe in God, do not believe in Good, do not love and are immoral evil hypocrites who use religion and/God for their own vain benefit in terms of power/fame/money at the expense of others.

The Unity of The Holy Trinity of **loving right action in truth, based on all the truth**—based on all known scientific, Biblical and experiential evidence as is described in this book —is the only way that your conscience-thoughts-emotions-body are straightly aligned and therefore is the only way that you are in an eternally unbreakable wholeness, oneness, a god. As a unique god you are in a Victorious State under any and all circumstances whether you succeed or fail.

Anyone who is not in this particular straight alignment of spirit and body **by righteous action**, finds conflict and brokenness within them, and finds external conflict also, so as a result they are on their way to at least some torment, angst, ultimately death and suffering after death if not Saved by The Christ Jesus.

Now, over the long term and eternally what is best for you is doing good, is righteous action to the right others for and in the right context of these new Independent Demilitarized Disarmed Democratic Holy States as seen by the eyes of God, who is the only One who ultimately truly matters.

The Christ Jesus knew and rightly prophesied that He would need to Manifest His Full Presence again, in Spirit this time, as the Son of Man. He knew that even after His First Presence in body the Jews who had been and are Delivered out of the tyrannical Egypt, would still have not been Saved out of their sin and death.
And He knew that even though the Gentiles, through the Greeks, would be saved by their faith in Righteousness, Christ Jesus, and that their faith and testimony would be treated as if righteousness, they would still have not been Delivered out of the spiritual and physical Babylon.
Therefore He Knew and rightly prophesied what He would do, during His Second Presence, which is to Save the Jews; and to separate and Deliver His Saved and blessed gentiles into Holy Independent Demilitarized Disarmed Democratic States.

To be delivered from evil one needs to keep the right distances from it.
While Salvation is by God's Grace, Deliverance is by one's righteous actions and righteous life in the right Independent non-violent society.
Buddha said: "You yourselves should act to achieve your deliverance." (Dhammapada 276)

This is the right time for the (Apostolic) Churches, Synagogues, Mosques and Temples to either let women become clergy and at least Bishops in the current militarized systems or to appoint them as clergy and Bishops for the small Independent Demilitarized Disarmed Democratic Holy State within your nation.

Be in Oneness and uniqueness by righteous action; by action that benefits others, truly, not your-self.

Did you get enough from the Mystic "Wine and Bread" of Life? Did this book guide you into "all of the truth?"

This is the Second Presence of the Christ Jesus.

God, in His Greatest Identity, the Father, will manifest Himself in about a thousand years and He will honor and justify me.

Do you believe in God? Do you truly believe in Jesus? Do you believe in Right?

Remember John Kennedy's words: "Here on earth God's work must truly be our own."

If yes, stop talking hypocritically about believing in right while acting wrongly, and establish the Kingdom of God, "on earth as it is in Heaven" (Matthew 6:10) by establishing at least one small Independent Demilitarized Disarmed Democratic Holy Land and State, within your nation, led mostly by women, where God's blessed may have the option to live truly free in Joy while in Peace.

If no, please stop hypocritically talking about democracy, liberty and freedom or about being right while enslaving and tormenting the multitudes of the disenfranchised, the poor, weak, depressed, pure in heart, pacifists and the righteous who do believe, and let them go and be also free, into a small, mostly women led, Independent Demilitarized Disarmed Democratic cooperation based State, for your own sake, to rebalance and realign your violent deceitful, hypocritical tormented nation.

If you don't do so fast, very fast, all hell will break loose as it has been written in Revelation. You have been very clearly warned.

Whatever you believe, the realistic/pragmatic choice in the Middle East is to either establish one Independent verifiably Demilitarized Disarmed State sanctioned by the UN somewhere there, and the ISIL held territory is the first obvious choice and to make it work or to let all hell continue to keep breaking loose.

Watch me go fast through Scylla and Charybdis as per Homer. Having done it so many times I now do it for sport: Dear wonderful, brilliantly shinning, more brilliant than the sun, the moon and the stars who can't see yet you can see, and can even see spiritually and understand a foreigner, an alien, whom you have never met or physically seen nor his image, through a 2 Dimensional virtual reality— accessed whenever you choose it — you, kind wise honest compassionate merciful loving fair amazingly courageous surprisingly strong stunningly lovely impressively artistic shockingly super-sexy excellently talented exceedingly intelligent honorable adorable uniquely and exceptionally beautiful friend: isn't now the right time for you to ask for and to do what is right for others and you and in doing so become truly Free?

We are on the clear; is everyone still with me?
At least "like" this!

Other than "Ithaca" being in shambles by the time you get there, the road is clear from here on, so be joyful and have the correct amount fun on the road, while you can!

Do what is right for others and live free eternally happy, in Joy while also being in Peace.
Please do and keep doing what is right... Thank you.

Some One: Son of Man.
Polydoros Andoni Zecos
I exist. I am
The Spirit of Truth

16. EXHIBIT

Unified Physics Theory of Everything

Deterministic Gravity and the electro-chemical and magnetic Quantum force are well known; the short-range Strong force is a Quantum force that holds the nucleus together despite the electromagnetic force's repulsion among protons; and is also countered by the short-range Weak (nuclear) Quantum force that causes decay though radiation to all fermions.

Given that solids have only one direction of motion available and that motion is determined by time, there is no "freedom of choice" as to their direction. In the liquid, gas, plasma States because motion in more than one direction at the same time is possible and likely, the laws (equations) governing those motions of matter are probabilistic.

This explains why General Relativity (GR) and/or Newton's laws that describe gravity of solid masses are deterministic, while the thermodynamics that describe the motions in the other 3 States of matter and Quantum Theory (QT) that describes the 3 Quantum forces and the motions of very small very fast particle-waves that are in a plasma, gaseous "cloud" state or liquid like State, are probabilistic.

We know that deterministic Gravity causes 1-directional motion of all 3 space Dimensions at a time so that spacetime becomes **4** Dimensional.

We know that the <u>local</u> $SU(3) \times SU(2) \times U(1)$ gauge symmetry is an <u>internal symmetry</u> that essentially defines the Standard Model.

And that the quantum, nuclear Weak force is described as having local **1** plus **2** Dimensional internal Symmetries; the quantum, nuclear Strong force as having local **3** D gauge Symmetry; and quantum Electromagnetism having a local **1** D Symmetry, of positive and negative charge.

And time is passive, in terms of these 3 quantum forces operating synchronously with the observer, showing their effect only when observed and at any time they are observed.

Therefore, if we add up all the dimensions of all the motions that happen at the same time, there are 11 Dimensional motions happening at a time.

Therefore, the 7 space dimensions, beyond the known 3 space dimensions and time, that the 11 Dimensional Super-Symmetry of the "M Theory" asserts, are not "curled up" undetectable and not measurable dimensions as has been theorized until now, which renders them unscientific but are the: 1 D plus 1, 2 D plus 3 D =7 synchronous but in opposition to gravity motions of plasma, liquids and gases "within" the 3D known space.

In the context of humans, these less than 4 dimensional motions— like those happening by gravity to our bodies— are the motions of our thoughts, emotions, words and those of "virtual reality" through our 2 D screens that happen at the same time as the 4-D gravitational motions but are also independent of them.

It is these additional to the gravitational motions that happen synchronously in a 1 dimension at a time, such as those of rays of light; and in 1, 2 space Dimensions at any time that they are observed as those of virtual reality; and 3D motions of and through air —such as during speech— that constitute the additional 7 space dimensions needed to find Supersymmetry in the Universe.

This would provide an understandable, reasonable, provable explanation for the otherwise not appearing realistic and not provable 10 or 11 Dimensions that

Super-Symmetry and string 'Theories of Everything' suggest.

So, Gravity and the solids it affects is deterministic; each and the sum of the other 3 forces, which are quantum forces is probabilistic; what is their relationship?

A force field is a function of the number of space dimensions occupied by the objects or particle waves by which that force interacts.

We define z as the number of the space dimensions occupied by the object or particle wave whose properties are being measured and call it the 'space dimensionality'—a term distinct from the term dimensionality as currently used to mean the units of measurement that are ultimately reduced to length, time, mass, charge— of the object.

-Objects that are 3 dimensional (3D) (space dimensions they occupy) have space dimensionality $z \equiv 3$

and particle-waves that are 0 dimensional points (0 D) have $z \equiv 0$.

Therefore, one may assume all the aspects of Riemannian 4D spacetime and General Relativity, GR, as correct for 3D objects; and one may also assume that the Hilbert space and the Quantum Theory, QT, as correct for 0 D particle-waves.

The rules allowing this mathematical TOE to identify specific mathematical objects for possible measurements are:

-When $z = 3$, GR, its equations (EFE equations) and Riemannian 4D space-time apply; and QT and Hilbert space geometry does not apply;

- When $z = 0$, Hilbert space geometry and QT and its equations apply; and GR and Riemannian geometry does not apply.

This way, GR and QT combined become a Unified Theory. Yet, that is not enough.

General Relativity (GR) and Quantum Theory (QT) are two limiting cases for 0 dimensional particle waves and 3 dimensional objects of a deeper mathematical theory that also allows for different theories to accurately describe objects or particle waves that have a space dimensionality between 0 and 3D. For example, string theories assume that the fundamental sub-particles are "one-dimensional curves called strings."

As a result, the mathematical rules are expanded to include string and super-symmetry theories, the Theory of Chaos and potential new theories, as follows:

-If and when $z=1$ then the string theory applies.

-If and when $z=2$ then the "M Theory," membranes and/or "strings and their dual particles," and super-symmetry in 11 Dimensions applies.

-If and when $0 < z < 3$ and is a fractional dimension then the Theory of Chaos (TC) applies.

So that all 4 basic Theories of physics: General Relativity; Quantum Theory; The Theory of Chaos; and Symmetry can be integrated.

Because of the first law of thermodynamics the change of Gravitational energy in any closed system including the Universe is equal to the sum of the change of energy in the 3 quantum forces, shown by the equation:

$$- \Sigma^z_0 \, \Delta E(z) . \, a_z = G \, \Delta s_z \qquad (1) \qquad 0 \leq z \leq 3$$

$\Delta E(z)$ is the change in the energy from the three Quantum forces as measured or as calculated by QT through the 1st, 2nd, and 3rd derivative Schrödinger equations, and/or through the Standard Model.

α_z as the a1, a2 and a3 are the already known effective coupling constants corresponding to each of the three quantum forces i.e., a1 for electromagnetism, a2 for the Weak, and a3 for the Strong force.G is Gravity as measured or as calculated by General Relativity or by Newton's equations. Δs_z is the distance traveled by the object with z space dimensionality.

Equation (1), and only this equation, gives the relationship between the four known forces of the Universe, as to the impact of changes of energy of the gravitational force to the changes in energy of the 3 quantum forces and vice versa.

As a subset of this equation (1), for $a_{0=}$ -1, G= m. c/dt, Einstein's equation can be derived: E=m c/dt . c. dt =m.c^2.

It is because of the opposition as above between gravity and the 3 quantum forces that unlike the equations for gravity, Schrödinger's equations for the 3 quantum forces, that describe images, have as a factor the imaginary, $\sqrt{-1}$ i.e., i.

As we know from the Theory of Chaos complexity and self-repetition arise because things have both a real and an imaginary component.

From equation (1) we can see that the way to reconcile deterministic vs. probabilistic outcomes, other than by the State of matter, is by the sum of all the probabilities of change in energy from the 3 quantum forces which equals to the negative of the change in energy in deterministic mass and its gravity.

As shown above there are sub-crosses on each axis of the Cross but not at random distances but at pre-defined distances, predefined orbits.

The 11 Dimensions— needed to get to Super-symmetry— that move at the same time are: a 4 Dimensional motion by gravity of solid objects; and at the same time, in reaction to it and causing change to that motion, 7 dimensional motions, as above, i.e. 3D+ (2 +1) D + 1 D= 7 D, by the 3 quantum forces, of and/or through the other 3 states of matter, (in simple language,) by and through air, water and fire; as seen through light.

The generalization of the above for life is:

The 11 Dimensions are: a 4 Dimensional motion by gravity of solid objects and our bodies; and at the same time and in reaction to and causing change to that motion, 7 dimensional motions, explained above, (and shown on the dimensional motions of the top 3 axes of the Cross) by our intuition, thoughts and emotions, through the 3 quantum forces and through the other 3 states of matter (in simple language) of and through fire, heat, pulse, water, and air, as seen through the light.

So, the importance of synthesizing one's understanding of the 12 fields of knowledge (Academic Disciplines), as it's done in this book, is that because each of these fields is a significant independent dimension of one's own existence, their synthesis is necessary for the synthesis of both Self, and of Everything.

SELECT BIBLIOGRAPHY

Recommended 100 Books

I am grateful to the authors of the following books:

1. *THE HOLY BIBLE* (from the Greek Biblio meaning Book,) any version.

2. *What the Buddha Taught* , by Walpola Rahula.

3. *Bhagavad-Gita*: As it is; any Edition.

4. *The Way of Life*; Lao Tzu; Translation by R. B. Blakney.

5. *The Essential Koran*; any translation, such as by Thomas Cleary.

6. *The Wisdom of Confucius*; Edited by Lin Yutang.

7. *The Odyssey* by Homer, any translation.

Religions

8. *The Complete Gospels* Edited by Robert Miller.

9. *Kabbalah* by Gershom Scholem.

10. *Kabbalah* by Perle Epstein.

11. *The Tao of Jesus* by John B. Butcher.

12. *Rabbi Jesus* by Bruce Chilton.

13. *The Holy Science* by Swami Sri Yukteswar.

14. *Tablets of Bahaullah* by a Committee; the Bahai World Center.

15. *The Prophet* by Kahlil Gibran.

16. *Understanding Islam* by Thomas Lippman.

17. *Great Christian Thinkers* by Hans Kung.

18. *The Imitation of Christ* by Thomas Kempis.

19. *The Spiral of Life* by Mona Rolfe.

20. *The Power of Kabbalah* by Yehuda Berg.

21. *The World's Religions* Edited by Sir Norman Anderson.

22. *The World's Religions* by Huston Smith.

23. *Our Religions* Edited by Arvind Sharma.

24. *Torchbearers of Spiritualism* by Mrs. St. Clair Stobart.

25. *The Religious leaders of Greece* by James Adam.

26. *Mere Christianity* by C.S. Lewis.

27. *The Seven Story Mountain* by Thomas Morton.

28. *Tai Chi Classics* translated by Waysun Liao.

29. *Buddhism* by Rhys Davids.

30. *Krishna the Charioteer* by Mohini M. Dhar.

31. *The Orthodox Way* by Bishop Kallistos Ware.

32. *Crossing the threshold of Hope* by John Paul II.

33. *The Power of Compassion* by the Dalai Lama.

34. *The Three-Personed God* by William J. Hill.

35. *Unconditional Life* by Deepak Chopra.

36. *The Great Thoughts* by George Seldes.

37. *The Lost Teachings of Jesus* by M. I. Prophet and E. C. Prophet.

38. *A Course in Miracles* by the foundation of Inner Peace.

Archaeology

39. *Archaeology of the Lands of the Bible* by Amihai Mazar.

Physics

40. *Modern Physics* by R. Serway, C. Moses, C. Moyer.

41. *THE NEW PHYSICS* Edited by Paul Davies.

42. *Quantum Mechanics* by F. Mandl.

43. *Relativity* by Albert Einstein.

44. *Space Time Matter* by Herman Weyl.

45. *Experimental Foundations of Particle Physics* by R. Cahn and G. Goldhaber.

46. *A Brief History of Time* by Stephen Hawking.

47. *The Feynman Lectures on Physics* by Richard Feynman, Leighton, Sands.

48. *Superstrings* Edited by Paul Davies.

49. *Chaos* by James Gleick.

50. *Fractals and Disordered Systems* by A. Bunde and S. Havlin.

51. *Theories of Everything* by John D. Barrow.

52. *Physics and Philosophy* by Werner Heisenberg.

53. *The Problems of Mathematics* by Ian Stewart.

Biology

54. *Descent of Man* by C. Darwin.

55. *Mapping our Genes* by Lois Wingerson.

56. *The Human Nervous System* by Murray L. Barr and J. A. Kiernan.

Fiction

57. *Hamlet; Macbeth;* and *King Lear*; by W. Shakespeare.

58. *Oedipus* plays by Sophocles translated by Paul Roche;

   Antigone, by Sophocles, any translation

59. *Lysistrata* by Aristophanes by Douglas Parker.

60. *Candide; Treatise on Tolerance;* by Voltaire.

61. *The Old Man and the Sea; A Farwell to Arms;* by E. Hemingway.

62. *The Celestine Prophecy* by James Redfield.

Philosophy

63. *The Portable Plato* Edited by Scott Buchman.

64. *The Basic Works of Aristotle*, any translation.

65. *From Socrates to Sartre* by T.Z. Lavine.

66. *The Story of Philosophy* by Will Durant.

67. *Making Sense of it all* by Thomas V. Morris.

68. *Greek Thinkers* by Gompertz translated by G. G. Barry.

69. *Self-Reliance and other Essays* by R. W. Emerson.

70. *Fuzzy Logic* by D. McNeill and P. Freiberger.

71. *English Etymology* by T.F. Hoad.

72. *The Philosophy Behind Physics* by Springer-Verlag.

Psychology/Psychiatry

73. *Diagnostic and Statistical Manual of Mental Disorders* by the American Psychiatric Association; latest Revision.

74. *Synchronicity* by C. G. Jung.

75. *Quantum Reality* by Nick Herbert.

76. *Quantum Psychology* by Robert Anton Wilson.

77. *Introduction to Neuropsychology* by J. G. Beaumont.

78. *Synopsis of Psychiatry* by H. J. Kaplan, and B. J. Sadock.

79. *Hierarchical Concepts in Psychoanalysis* Edited by A. Wilson and J. E. Gedo.

80. *Severe Personality Disorders* by Otto Kernberg.

81. *Of 2 Minds* by T.M. Luhrmann.

82. *Cultural Psychology* by J. W. Stigler, R. A. Shweder and G. Herdt.

83. *Beyond Boredom and Anxiety* by M. Csikszentmihalyi.

84. *Intellect* by Mortimer J. Adler.

Economics

85. *The Wealth of Nations* by Adam Smith.

86. *The General Theory on Employment, Interest and Money* by J.M. Keynes.

87. *Capitalism and Freedom* by Milton Friedman.

88. *The Politics of International Economic Relations* by Joan Edelman Spero.

89. *Land without Justice* by Milovan Djilas.

90. *Individual Rights in the Corporation* by A. F. Westin and S. S. Salisbury.

Politics

91. *The Declaration of Independence; US Constitution; Federalist Papers.*

92. *Democracy in America* by Alexis de Tocqueville.

93. *World Politics* by Bruce Russett and Harvey Starr.

94. *Mandate for Peace* by M. Gorbachev.

Organizations

95. *Managing Organizational Behavior* by Cyrus F. Gibson.

96. *Behavior in Organizations* by A. G. Athos and R.E Coffey.

97. *Mastering Change* by Ichak Adizes.

98. *The Effective Executive* by Peter Drucker.

99. *The World Anew; The Book of Life; The Book of Truth* by Paul Zecos!

100. *Walt Disney's Donald Duck* by Carl Barks!

www.ingramcontent.com/pod-product-compliance
Lightning Source LLC
Chambersburg PA
CBHW051858170526
45168CB00001B/150